BATHROOMS

ARCHITECTURAL HANDBOOKS SERIES

BATHROOMS

Charanjit S. Shah

B Arch, FIIA

Practising Architect

Chairman, The Indian Institute of Architects, Northern Chapter (1994-96 and 1998-2000)

Professional Advisor, Architect of the Year Award 1997

Visiting Professor, School of Planning and Architecture, New Delhi

CBS PUBLISHERS & DISTRIBUTORS PVT. LTD.

NEW DELHI • BENGALURU • CHENNAI • KOCHI • MUMBAI • PUNE

ISBN: 81-239-0776-1

First Edition: 2002
Reprint: 2007, 2012, 2015

Published by:
Satish Kumar Jain for CBS Publishers & Distributors Pvt. Ltd.,
4819/XI Prahlad Street, 24 Ansari Road, Daryaganj, New Delhi - 110002
delhi@cbspd.com, cbspubs@airtelmail.in • www.cbspd.com
Ph.: 23289259, 23266861, 23266867 • Fax: 011-23243014
Corporate Office: 204 FIE, Industrial Area, Patparganj, Delhi - 110 092
Ph: 49344934 • Fax: 011-49344935
E-mail: publishing@cbspd.com • publicity@cbspd.com

Branches:
• *Bengaluru:* 2975, 17th Cross, K.R. Road, Bansankari 2nd Stage, Bengaluru - 70
   Ph: +91-80-26771678/79 • Fax: +91-80-26771680
   E-mail: cbsbng@gmail.com, bangalore@cbspd.com
• *Chennai:* No. 7, Subbaraya Street, Shenoy Nagar, Chennai - 600030
   Ph: +91-44-26681266, 26680620 • Fax: +91-44-42032115
   E-mail: chennai@cbspd.com
• *Kochi:* 36/14, Kalluvilakam, Lissie Hospital Road, Kochi - 682018
   Ph: +91-484-4059061-65 • Fax: +91-484-4059065
   E-mail: cochin@cbspd.com
• *Mumbai:* 83-C, Dr. E. Moses Road, Worli, Mumbai - 400018
   Ph: +91-9833017933, 022-24902340/41 • E-mail: mumbai@cbspd.com
• *Pune:* Bhuruk Prestige, Sr. No. 52/12/2+1+3/2,
   Narhe, Haveli (Near Katraj-Dehu Road Bypass), Pune - 411041
   Ph: +91-20-64704058/59, 32342277 • E-mail: pune@cbspd.com

Representatives:

• Hyderabad: 0-9885175004   • Kolkata: 0-9831437309, 0-9051152362
• Nagpur: 0-9021734563   • Patna: 0-9334159340
• Vijayawada: 0-9000660880

Printed at:
J.S. Offset Printers, Delhi

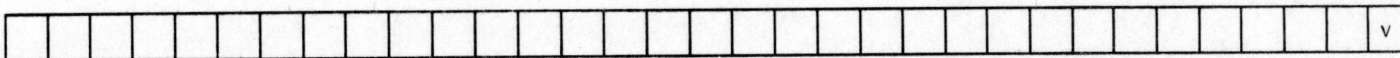

Foreword

Plumbing practices in our country are equally varied, modern and traditional.

In a present day building, the plumbing practice has to be the most updated, well-worked out and well executed. Good practice starts from the drawing board. It is at this stage, that the designer needs a single source of comprehensive information for the total design and engineering plumbing. It is for the first time, in my knowledge, that one could find such comprehensive information of practices, materials, systems, fittings and fixtures, and everything else associated with plumbing aspects and bathroom as a whole in one volume. Ar Charanjit S. Shah has accomplished this task by putting within one cover almost all that a designer would need to know on preliminary designs and engineering of plumbing for day to day use in housing and small scale industries. This book should be an equally useful source of information for a student of the subject and a student of architecture in particular.

It is interesting to read the history of bathroom design and development of fixture over a period.

What would be the toilet of future? Let us wait and see.

Prof PK Chowdhry
Ex-Head, Department of Architecture
School of Planning and Architecture
New Delhi

to

my son
Prabhpreet Singh Shah

student of M.Tech in Information Technology, who is an upright,
inspiring personality always keen and eager to help. With his
sincere and dedicated architectural spirit and pushing attitude that this
dry, rough and boring subject has been made lively and pleasant.

He has the zeal, dedicated spirit, encouraging sentimental approach,
understanding and in-depth knowledge with a noble and humble approach in life.

Preface

It is generally said that architecture is the mother of all arts. To create good architecture, it is necessary to know all the subjects related to building industry.

The book *Bathrooms* is yet another volume in the Architectural Handbooks Series that consists of various volumes on different architectural subjects. Two other volumes, *Architects Hand Book: A ready reckoner* and *Water Supply and Sanitation,* have already been published. The main purpose of this series is to provide everything a professional might need, at any stage of building execution.

Technical information regarding all aspects of architecture, with special emphasis on Indian conditions, has been included. These volumes shall cover detailed information needed at various stages of designing in a very practical and systematic manner.

This volume is another step towards more knowledge in architecture. It is a guide in simple formation to the professionals involved in designing and conceiving of building execution. Though bathroom is only a part of the house, it can cause a lot of discomfort to the users if not planned carefully. The book also presents the details of type of fittings and materials available and the norms and standards as related to the bathroom designing. It also covers various possible types of bathrooms, from the bare necessities to the "super luxury rooms."

All aspects of bathroom planning, colours, textures and lighting, etc. too have been dealt with. Storage and plumbing have also been discussed in detail. The product profile and the desired information have also been provided, keeping in view the affordability and sustainability of various materials.

Waterproofing, water heating systems by way of solar and electrical systems have been elaborated in the most professional way.

The space requirements and design concepts of public toilets have also been critically analysed with a definite purpose of awareness of space distribution, circulation, and behaviour of various materials. The understanding of various plumbing systems have also been highlighted.

This complete course on *Bathrooms* shall provide all technical knowledge one needs while designing a bathroom.

A need was felt for a single book that could include all the information, details and technical aspects of bathroom designing and planning. Hence, this book was born.

The architect's point of view has been properly elaborated giving detailed technical information on various aspects of bathroom designing.

Charanjit S. Shah
Practising Architect
Chairman, Indian Institute of Architects,
Northern Chapter (1994-96 and 1998-2000)
Professional Advisor
Architect of the Year Award 1997
Visiting Professor, School of Planning and
Architecture, New Delhi.

Acknowledgement

This publication *Bathrooms* is another volume of the architectural series *Architectural Handbooks Series*. It has become possible with the active moral support of my professional colleagues, fellow professionals, friends and certain institutions. I am indeed grateful to my following professional colleagues for their active participation and moral support.

Prof (Ar) MM Rana, Dean, Sushant School of Art and Architecture, Gurgaon

Ar Parmendra Raj Mehta, President, Council of Architecture, New Delhi

Dr (Ar) RS Sodhi, Practising Architect

Ar KB Mohapatra, Director, Ajay Binoy, Institute of Technology, Piloo Modi College of Architecture, Chairman, Indian Institute of Architects, Orissa Chapter

Ar Alok Ranjan, Head, Department of Architecture, Malviya Regional Engg College, Jaipur Vice-Chairman, Indian Institute of Architects, Rajasthan Chapter.

Ar Suresh Arora, Sr Architect, Ministry of Health, Nirman Bhawan, New Delhi.

I also express my gratitude to all those who have helped me in presenting the book in its final form. I am delighted to put on record the tremendous and sincere efforts put in by the young Architect Savita C Sengar and Gurpreet S. Shah to make this a sucessfully represented volume in the *Series*.

I am deeply impressed by the most professional work of Mr YN Arjuna in compiling and producing this book. My warmest thanks to Mr Satish Jain of CBS who has been instrumental in giving a practical shape to bring out the book.

[Charanjit S Shah]

Contents

INTRODUCTION

Although bathrooms made their appearance in the Egyptian palaces in the pre-Christian era, it were the Romans who gave it a pride of place and transformed bathing into a luxurious art. Exquisite bathhouses were built and attention paid to even the smallest of bathing ritual. The Roman baths offered facilities for steaming, massage, cleaning and even icy plunges.

In the early days, bathrooms in India were not within the four-walls but rivers served the purpose and such description is often found in Indian mythology like *gopies* taking bath in the river and Lord Krishna stealing their clothes.

The Greek preferred to take cold showers by pouring water over themselves and this tradition ultimately evolved into the public bath system.

Bathing became a social activity which the Romans developed into an art. But later on, it shifted to inside the house, as in the mid-century we find description of swimming pools for the queens.

In India, we had our own traditions and customs, bathing in rivers and pools was one of them. The British, introduced the concept of separate room—the washroom. In the last few decades due to the changing as well as advancing technology, there has been a major revolution in bathroom designing and fittings and evolution of luxurious five star bathrooms.

Bathroom, an area which in the homes was considered unnecessary earlier, has now become as important as the living room. More and more people are opting for more cheerful bathrooms, the affluent are looking for newer ways to be different, and that is why the architects and designers have begun to conceptualise the bathroom.

The reason for giving so much attention to bathrooms is that they are not just bathing places but are gaining a place of prestige and status. Now they are decorated with colourful tiles and many other decorative and luxurious items. Hence there is a need felt in elaborately discussing all pros and cons related to bathroom planning, its standards, trends in sanitary wares, bathroom fittings, fixtures, and most important of all, the maintenance.

A few years ago, bathroom was considered the most insignificant part of the house and was often relegated to the back of the house or the outside, and totally ignored at the time of planning. But now the concept of the bathrooms has totally changed. Since then bathrooms have gone into transformation. With growing consciousness and technical innovation, bathrooms have taken an important place in the house. The manufacturers are giving considerable attention to the new products. Gone are the days when there were only white sanitary fittings with limited design range to choose from. Today, market is full of a vast variety of materials, sanitaryware and fittings, that one often gets puzzled at what to buy. A comprehensive guide is needed where one can get the full information about the bathrooms, be it planning standards, criteria, design elements or materials range available in the market. In this book there is everything that a professional needs to know about bathrooms. This book covers all the aspects of bathroom planning and shows the different ways by which the maximum utilisation of the most functional rooms can be achieved. It covers the minimum requirements of the bathrooms to the most luxurious items, with minimum space standards.

This book covers:

- basic bathroom standards and requirements

- various layouts for the large and luxurious bathrooms to the smallest of the utility bathrooms

- plumbing and drainage network which is instrumental in making these rooms functional ones

- design elements which may change the whole look of the bathrooms and add life to it

- storage, ventilation, heating and sound insulation of the bathrooms

- materials and designs of the sanitaryware and fittings as available in the market

- maintenance tips and other problem areas in the bathroom which, if not given proper attention, can lead to a lot of inconvenience

This book is a result of more than three decades of the author's experience for planners and the users. This book is a guide and source of information and of reference to stimulate new ideas. The result will not be just a functional room but much more than that.

The same have been described in this book starting from history of bathrooms to the evolution and luxurious bathrooms and use of new innovative and revolutionary fittings and faucets.

Ar Charanjit S Shah

HISTORY OF BATHROOMS

Bathrooms have gone into transformations from the ancient times to the present age. From earliest times, man has designated spaces for the purpose of bathing and washing which were subsequently known as bathrooms. The different ages have responded to their hygiene needs in different ways and with varying levels of sophistication, governed largely by prevailing prejudice or religious and social attitudes. Attitudes to bathing vary interestingly across the world.

Indus valley excavations have shown vestiges of a large bath or tank in Mohenjodaro with a range of cells, the earliest well-preserved examples of bathrooms belong to the Aegean civilisation, C. 1700-1400 BC. These are notable for their careful structure as well as their advanced system of water supply and drainage. The earliest bathrooms were recorded in Cretan city of Knosses where the Minoans had a high standard of hygiene, and many homes had their own bathrooms. The Greeks preferred to take cold showers by pouring water over themselves, and this tradition ultimately evolved into the public bath system. Ancient vase paintings depicting showers and public baths highlight the importance given to bathing in the life of the Greeks, but they do not display any special or monumental architectural treatment.

It was the Romans, with their love for luxury, who developed an elaborate technique of bathing and refined it into a fine art. Elaborate bath-houses were planned and built with exquisite attention paid to every aspects of the bathing ritual. This elaborate ritual was undertaken in baths of varying sizes, from private homes where each unit was merely a small room, to the enormous thermae of imperial households. The arrangement of rooms in the smaller baths was fixable and usually consisted only of the essential rooms. The imperial thermae also served as a great social centre with gardens, a stadium and pavilions where lectures were given and poems read.

The general scheme comprised a great open garden surrounded by subsidiary club rooms, and a block of bath chambers either in the centre of the garden, or at its rear. The enormous rooms were roofed and illuminated by an ingenious system of buttressing, cross-vaulting and clerestory windows. The splendour of these bathing establishments was enhanced with rich furnishings, sculpture and architectural decoration. Bathers padded across marble or mosaic floors that were heated from below, while slaves scurried through underground passageways to provide smooth and efficient service without being seen. The walls too, were clad in marble and decorated with stucco reliefs, colour and mosaic. Gilt bronze was freely used for doors, capitals and window screens. Men and women bathed separately in the Roman baths, and usually at different times or in different establishments. Mixed bathing was not generally approved. The extreme luxury of Roman public baths later came to an end with the fall of Roman empire.

At one time, the bath was something of a social event. However, by the fourteenth and fifteenth centuries, bathing became increasingly promiscuous, the best traditions of Roman custom continued in Arabian Islamic countries and has survived to the present day in the form of the Turkish hammam which is a public bathing house where people undergo the entire ritual of bathing on similar line to that of the ancient Romans. The closest equivalent today would probably be the five star heath club or spa. After the fifteenth century, the reformation and the rise of puritanism, the bathing was considered a sin. Genteel folk disapproved of bathing as being utterly depraved. Besides, middle class homes rarely had bathrooms. People washed in front of the living room fire, sometimes in curiously shaped baths like that of a boot, where the bather sat with his or her head sticking out of the ankle. However this remained only an occasional exercise. In an age when baths were not accepted in polite society, men and women took steam baths together.

In Japan, the classical bathrooms were of wood with sloping floors to allow the extra water to drain out. Baths traditionally consisted of large, wooden tubs shared by the whole family using the same filling of extremely hot water.

In Russia, simple bath-houses consisted only of a steam room and cold bath. For the Indian, more specifically the Hindu, bathing is not merely body cleansing, but is associated with the purity. Hindus are well known for their obsession with keeping bodies clean probably owing to the hot, sticky, tropical climate. The Mughals considered bathing an important activity and built numerous hammams where their *begums* had long luxuriously, scented baths with the help of several attendants. In India, we had our own tradition of bathing in the river or at best by the well in the open, until British Victorian influence introduced the tin tub or wooden commode to our homes. However, even today, the villager at least still prefers the open fields and regularly visits his or her privy at daybreak with the *lota*. Alternatively, in urban settlements where privacy could not be obtained in the open, lavatories were located in outhouses, where one squatted on the floor, with the feet firmly planted on footrests on either side of a hole on the ground, or a tin tray. The tray was emptied regularly manually. The unhygienic form of waste disposal was fairly widespread before the days of modern sanitation and is known as conservancy system. This traditional system gave birth to the Indian style toilets which were always situated a little away from home to keep the house free from foul odours. But it was very inconvenient to use such bathrooms at midnight or during rains.

There used to be a separate space for toilets and bathing areas. Since bathing assumes such significance, special attention was paid to the construction of the bathroom. There were provisions to heat water in the bathroom. It was so constructed that a person could stand outside the bathroom and start a fire under the vessel which was kept inside the bathroom. In affluent houses, the bathrooms had cement tanks which could be filled from the outside with water drawn from the well. The floors were kept rough to prevent it from becoming slippery from the oils used for the bath.

By the end of nineteenth century, things began looking up with the introduction of city water supply and sewage system in London. The valve water closet, a proper flushing lavatory was developed by Joseph Braham. Shower baths also began to make its appearance. Thus, bathroom gained its importance and began to be accepted as an integral part of the house. It was only then that the bathrooms moved indoors. Health and privacy concerns, not aesthetics prompted the move. Today, gone are the Spartan dingy areas where ablutions were performed as a necessity and comfort was totally absent. The bathroom of today is a kind of place, somewhere you can escape far from the maddening crowed, to relax and recover your equilibrium. In the last couple of decades, due to changing attitudes as well as advancing technology, a revolution has taken place in bathroom design, and an area of home that received the minimum attention is now enjoying a revival. Modern bathrooms wear a well-dressed look, its design recognising people's needs to begin and end their day in pleasant surroundings. It is after all an intimate personal place—perhaps the only room in the house where one can be totally private and get much needed peace. Bathroom means a place which is bold, beautiful and especially comfortable.

The new bathrooms are bigger and compartmentalised for multiple uses, and many are geared for relaxation as well as efficiency. Exercise rooms, saunas, steam showers, grooming alcoves, walk in dressing wings, even indoor atriums are all new master suite options. Apart from cleaning the body and other intimate rituals, it is a place to revive, freshen up, read, exercise, plan the day ahead or just switch off and unwind.

Finally, bathroom remains the most important room in the house. It is the one facility in home that each one of us uses every single day of our lives. Everything in it should, therefore, be functionally perfect, sanitary and easily maintained. Depending on the personal preferences, the rest is all icing on the cake.

PLUMBING AND DRAINAGE

The network of pipes that brings water to the taps and takes away the waste and soil is known as plumbing. The plumbing of the building starts and ends outside the building with the town or city's water supply and drainage system. The plumbing drainage system plays a very important role in the layout of the bathrooms. The plumbing system of the bathrooms can be divided into two parts:

(i) Water supply
(ii) Waste and soil drainage

WATER SUPPLY

Water is collected from its sources, and taken to the treatment plants for the proper treatment. The treated water is then transmitted to the service reservoirs, serving the town or city. Municipality is responsible for the distribution of water from the service reservoir to the user. The distribution network can be divided into two levels:

(i) At municipal level
(ii) At domestic level.

At Municipal Level

This includes the network of trunk mains and street mains. Municipality is responsible for the distribution of water from the service reservoir till it reaches the street mains. Designing of this network involves a very lengthy procedure and is out of scope of this book, and only broad aspects have been discussed here.

Systems of Supply

There are two systems of supply of water from the mains to the service pipes depending upon the availability of water:

(i) Continuous system
(ii) Intermittent system.

In continuous system, the water is supplied to the consumers 24 hours a day. This system is definitely preferred but is not always possible due to poor water pressure availability and insufficient quantity of water. In intermittent system of water supply, water is supplied to the consumers for certain fixed hours in a day, usually about 1 to 2 hours in the morning and same in the evening. In this system, the area is divided into several zones, and timings in every zone is adjusted to maintain a proper working pressure.

Intermittent system has certain drawbacks, like it demands greater sizes of water mains as the supply is given for limited number of hours. Second, it is very difficult to fight fire, in case it breaks out during non-supply hours.

At Domestic Level

At this level, water is conveyed from the street mains to the individual building, and then to the taps and other fixtures, etc. The supply from the street main to the individual is made through house service connections.

The house service connection consists of two parts:

(i) The communication pipe which runs from the street mains to the boundary of the premises.
(ii) The service pipes which run inside the premises. The communication pipe is laid and maintained by the local authority at the cost of owner of the premises, while the service pipe is laid by the consumer at his or her cost. To prevent damage by road traffic, communication pipe must be placed 0.76 m, below ground level and the depth is maintained until it is inside the building. The rising pipe is brought up on internal wall rather than external wall where it can be affected by weather conditions. A stop valve is placed immediately above ground floor level. In the planning of the building, the need is to centralise the plumbing to give an efficient and economical layout of the appliances requiring water, i.e. bathrooms close to the kitchen, and water tank above the bathrooms, etc.

Systems of Supply

The water supply from the mains to the building is through one of the following systems depending on the pressure of water in the street mains and the timings of supply. Sometimes, both the systems are used for supply of water.

Direct supply system (Upward distribution system) The supply of water is given directly to various floors from the water mains having sufficient pressure for sufficient number of hours. This system is recommended only if the number of floors in the building is not more than two. Separate connections for domestic and non-domestic requirements are provided.

Indirect supply system (Down take supply system) When the water pressure in the mains is not sufficient for the direct supply, water from the mains can be either:

(i) Pumped up into the overhead storage tank, usually situated at the roof of the building from the mains.
(ii) Stored into the underground storage tank from where the water is pumped to the overhead tank and then it is supplied by gravity, or
(iii) both.

In indirect supply system, booster pumps are used to supply water to the overhead storage tanks. The suction pipe of the booster pump should not be connected to the distribution main because this will affect the water supply of the whole neighbourhood, specially where water pressure is low. The best method is to fill overhead tanks by boosting water up from sump, a covered tank at ground level. The sump, well protected from possible contamination may be connected and fed directly by the service pipe, and a small electric pump raises the water from sump to overhead tank.

To avoid the overflow of water in the overhead tank, automatic pump controllers are very effective. These automatically switch off the pump when water level in the overhead tank rises above the required level. Pump controller also prevents dry running of the pump.

Automatic Motor Pump Controller

Sometimes, using a motor pump also does not solve the problem of regular water supply completely. In fact, new set of problems arise in the form of everyday operation of the motor pump. The daily operation of the water pump involves a number of jobs. These jobs begin every morning with the checking of water level in this overhead tank, then switching on the water pump, then waiting for a long time till the overhead tank overflows, and then rushing and switching off the motor pump. At times, there is no electricity when it is needed the most. The solution can be an automatic system so that the following could be avoided.

(i) Overflow from overhead tank and hence, wastage of water, electricity and damage of roof
(ii) Burning of motor due to dry running
(iii) Air locking of motor
(iv) Keeping in mind to switch on and switch off the motor
(v) Going up and down to check water level in the tanks

Functioning of Automatic Motor Pump Controller

Automatic motor pump controller is an electric system consisting of two integral parts:

(i) Autolift,
(ii) A pair of sensors.

Both these parts function simultaneously. Two sensors are fixed inside the tank to define two specific levels of water: one long-length sensor is used to define a lower level, whereas one short-length sensor is used to define an upper level. These sensors are connected to automatic controller through cable. At the same time, the main supply to the motor pump is monitored through autolift unit with the rise and fall of water level inside the tank. Signals are generated through sensors which are passed into the autolift unit, which in turn switches on or switches off the motor pump as required.

Applications

It has various applications in industries, distilleries, public institutions, multistoreyed buildings and apartments, chemical treatment plants besides the domestic use in houses.

These are four common applications involving the use of domestic motor pumps:

(i) Storage of municipal water into an underground/ground level tank and transferring this water into an overhead tank using a booster pump. Power controller starts the pump automatically when water level in the overhead tank falls below the lower (start) sensor. Simultaneously, the pump stops immediately if water level in the ground level tank fall below the lower (stop) sensor. Once stopped this way, pump controller restarts the pump only when water level in the ground level tank rises up and touches the upper (start) sensor (Fig. 3.1).

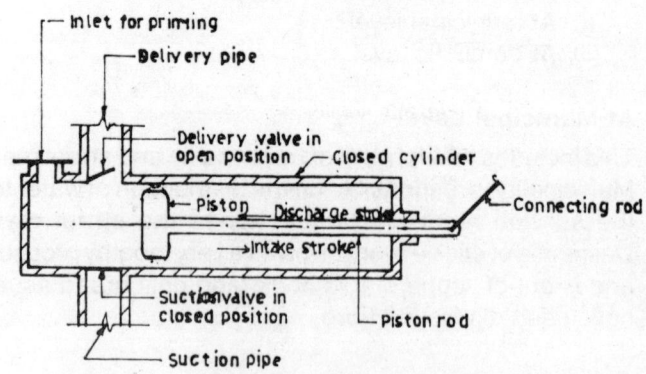

Fig. 3.1 *Reciprocating pump.*

(ii) Filling of an overhead tank with tubewell water using a tubewell/jet pump. Power controller starts the pump automatically when water level falls below the lower (start) sensor. Once started power controller keeps the pump running till water level touches the upper (stop) sensor (Fig. 3.2).
(iii) Collection of waste water of the building into an underground/ground level tank and disposing off this waste water into the sewage system using a sump pump. Power controller starts the pump automatically when water level touches the upper

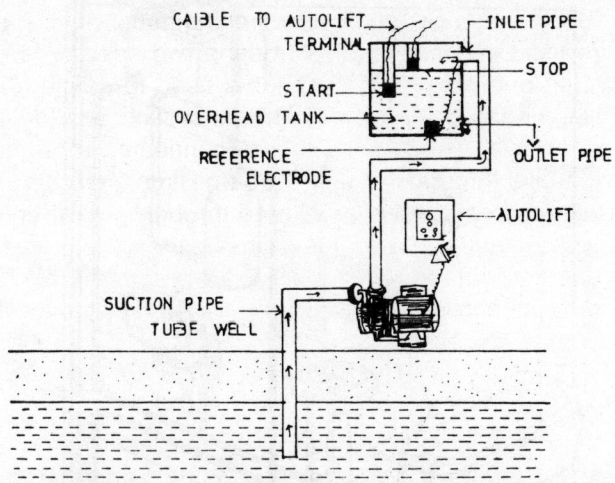

Fig. 3.2 Autolift with tube well/jet pump.

(start) sensor. Once started, power controller keeps the pump running till water level falls below the lower (stop) sensor (Fig. 3.3).

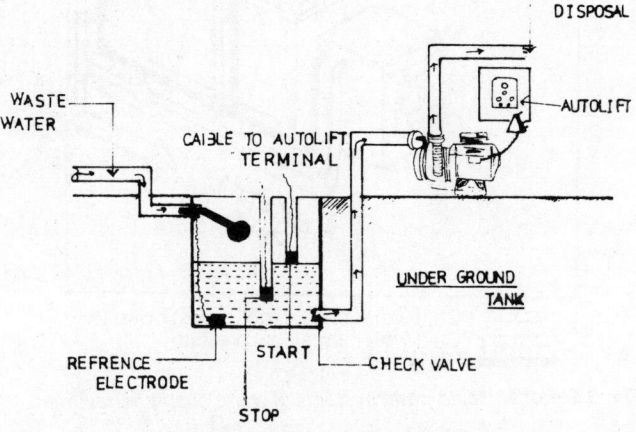

Fig. 3.3 Autolift with sump pump.

(iv) Filling of the overhead tank with municipal water using a "Direct on Line". This is a semi-automatic type. When water supply is available, the switch has to be started manually with the help of a push button. Once started, the power controller stops the pump automatically when water level touches the upper (stop) sensor (Fig. 3.4).

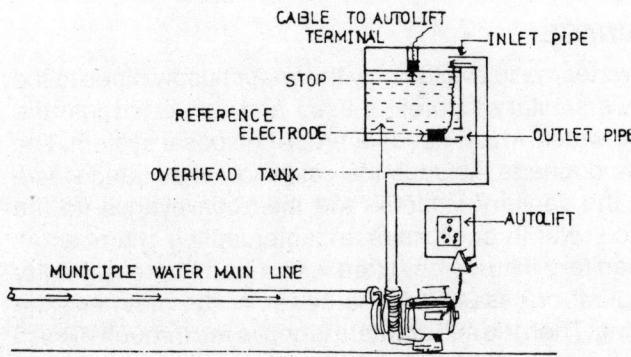

Fig. 3.4 Autolift with direct on-line pump.

Water Requirements

In indirect supply system, the provision of water storage tank is a must. Now, the point is what should be the capacity of water storage tanks for regular supply of water in the taps. To assess the capacity of water storage tank, it is important to know daily water requirement of a person for various activities (Table 3.1).

On the basis of the above requirements, it has been assumed that a person uses an average of 180 litres a day including the water required for flushing water closets.

Table 3.1: *Daily water requirement for various activities*

Activity	Quantity litres/head/day
Drinking	1
Cooking	5
Bathing	35
Washing hands/face, etc.	8
Household sanitary purposes (Washing clothes, utensils, etc.)	50

The underground storage tank should have a minimum storage capacity equivalent to 50 per cent of daily demand if water supply is intermittent. In such cases, the capacity of the overhead tank should be minimum of one day's requirement.

For domestic purposes, the storage requirements are calculated on the basis of basic requirement per tenement, subject to a certain minimum storage based on number of down take fittings such as taps, showers, bathtubs, etc. fed from the storage tanks.

Normally, separate flushing tanks are provided to supply water for flushing cisterns (for WC and urinals) in the building through downtake pipes. The storage capacity of flushing tank is calculated on the basis of number of WC seats installed in the buildings.

Table 3.2 shows the storage capacities for residential buildings. Sometimes, flushing tanks are provided separately for WCs

Table 3.2: *Storage capacity of water for residential buildings*

	Domestic Storage Capacity	Flushing Storage Capacity
A. For tenements having common convenience	500 litre per tenement	900 litre C. Seat
B. For residential premises occupied as flats or blocks	800 litre per tenement	270 litre net for each WC seats and 180 litre each additional seat in the same flat

The capacity of the water tank primarily depends on the water requirement of the building and on the hours of supply from the mains with pressure sufficient to fill up the overhead tank, if only overhead tank is provided.

The underground storage tank should have minimum storage capacity equivalent to 50 per cent of daily demand if the water supply is intermittent. In such cases, the capacity of the overhead tank should be maximum of one day requirement.

Water is supplied from the overhead tank to the tap through a ring system. The number of pipes directly connected to the storage tank should be minimum to maintain a sufficient water pressure. Thus, the network of pipes consists of one main delivery pipe which delivers water to the branch pipe. Water from the storage tanks is supplied to the taps through branch pipes. The maximum number of branch pipes that can be connected to the main delivery pipe depends on the diameter of the main delivery pipe as well as the diameter of the branch pipe. Table 3.3 gives the maximum number of branch connections which can be fed from the main delivery GI pipe for continuous water flow in bath/kitchen, e.g. if the diameter of the main delivery is 100 mm, and that of the branch pipe is 90 mm, only one branch connection is fixed, the diameters of branch pipes can be calculated from the Table 3.2, e.g. if 3 connections are to be made from a 80 mm main delivery pipe, the diameter of branch pipe should not be more than 50 mm. Thus, the above table helps in calculating the number of branch connections if the diameter of the main delivery GI pipe as well as the branch pipe is fixed, and vice versa.

Table 3.3: *Number of branch connections along with the diameter of pipes from the main delivery GI pipe*

Diameter of main delivery line (mm)	Diameter of branch pipe line (mm)									
	100	90	80	65	50	40	32	25	20	15
100	1	1	2	3	6	10	17	32	53	113
90	x	1	1	2	4	8	13	25	43	88
80	x	x	1	2	3	6	10	18	32	66
65	x	x	x	1	2	3	6	11	19	39
50	x	x	x	x	1	2	3	6	10	20
40	x	x	x	x	x	1	2	3	6	12
32	x	x	x	x	x	x	1	2	3	7
25	x	x	x	x	x	x	x	1	2	4
20	x	x	x	x	x	x	x	x	1	2
15	x	x	x	x	x	x	x	x	x	1

For Separate Hot and Cold Water Supply

In continuous system, the cold water supply is taken directly from the municipal mains, and the hot water tap is connected to the overhead tank through a geyser or any other type of heating system. In intermittent system, both the hot and cold water tap are connected to the overhead tank. The cold water tap is directly connected to the overhead tank, but the hot water supply is through an electric water or solar heater (Fig. 3.5).

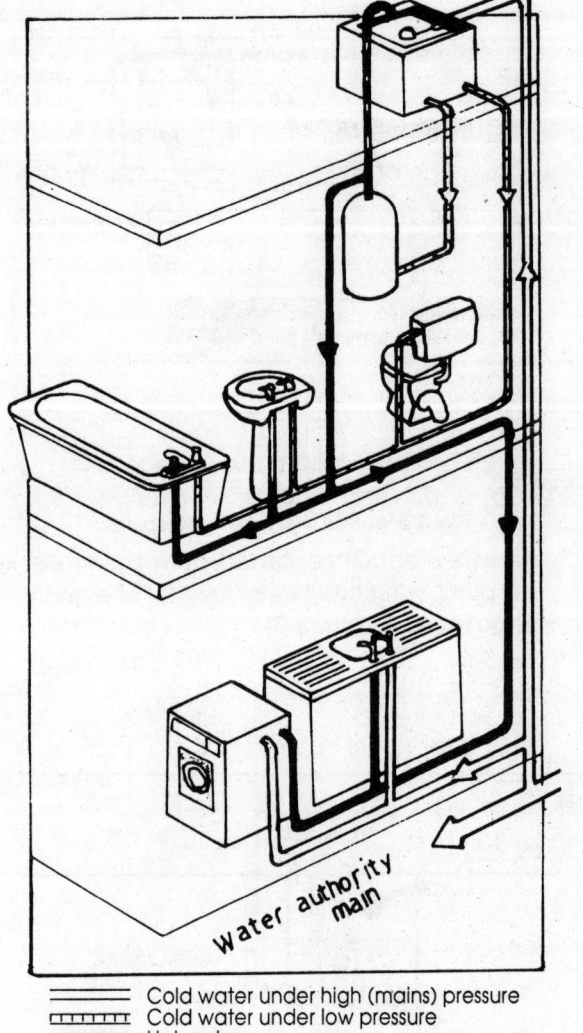

Cold water under high (mains) pressure
Cold water under low pressure
Hot water

Fig. 3.5 *Hot and cold water systems of water supply in bathroom.*

There are two types of heating systems: electric and the solar water heating system. Separate electric heaters of smaller capacities are provided for single bathrooms, but there is a centralised heating system as well as which would supply hot water to more than one bathrooms. In central water heating system, be it electric or solar, all the hot water pipes should be properly insulated.

DRAINAGE

The water made available by the water supply pipes to the various sanitary fixtures is used and converted into the waste water, which needs a proper disposal system. The house drainage includes the collection of soil and waste from the sanitary fixtures, and their conveyance, to the public sewer through traps and intercepting chamber. All the sanitary fixtures are fitted with a trap to prevent entry of foul air or gases from the sewer or the drain into the building. Then the soil or waste is conveyed through the soil pipe or waste pipe respectively to the house drain which in turn is connected to the inspection chamber. The house drain is connected to the municipal sewer through interception chamber, which separates the two.

Systems of Plumbing

For the drainage of waste and soil, four types of systems (Figs 3.6 to 3.8B) are usualy employed in building:

  (i)   One pipe system
  (ii) Two pipe system
  (iii) Partially ventilated single stack system
  (iv) Single stack system.

One Pipe System

In this system, only one set of pipes are used—main soil waste pipe and main vent pipe. All the contents of kitchen sinks, bathroom floors, washbasins and water closets are discharged into a single pipe. The soil waste pipe is directly connected to the building drain. Gulley trap is not provided. The vent pipe of minimum 50 mm diameter provides ventilation to water seals of all the traps. All traps should have a minimum 75 mm deep water seal. In multistoreyed buildings, the waste branch should join MSWP above the soil branch at each floor (Fig. 3.6).

Two pipe System

In this system, two sets of pipes are used for drainage:

  (i) Soil pipe with vent pipe
  (ii) Waste pipe with vent pipe.

All the contents of water closet (WC) and urinals are discharged into soil pipe. The waste pipe contains waste water from washbasins, sinks, bathroom floors, etc. The soil pipe is directly connected to the building drains whereas, the waste pipe is connected to the building drain through the gulley trap. Both the pipes are provided with separate vent pipes. Thus, this system contains four pipes (Fig. 3.7).

Partially Ventilated One Pipe System

In this system, single soil cum waste pipe carries the discharge from water closets, urinals, bathrooms, sinks, etc. It differs from one pipe system as the traps of soil fittings, i.e. water closets and urinals only are ventilated through single vent pipe (Fig. 3.8A).

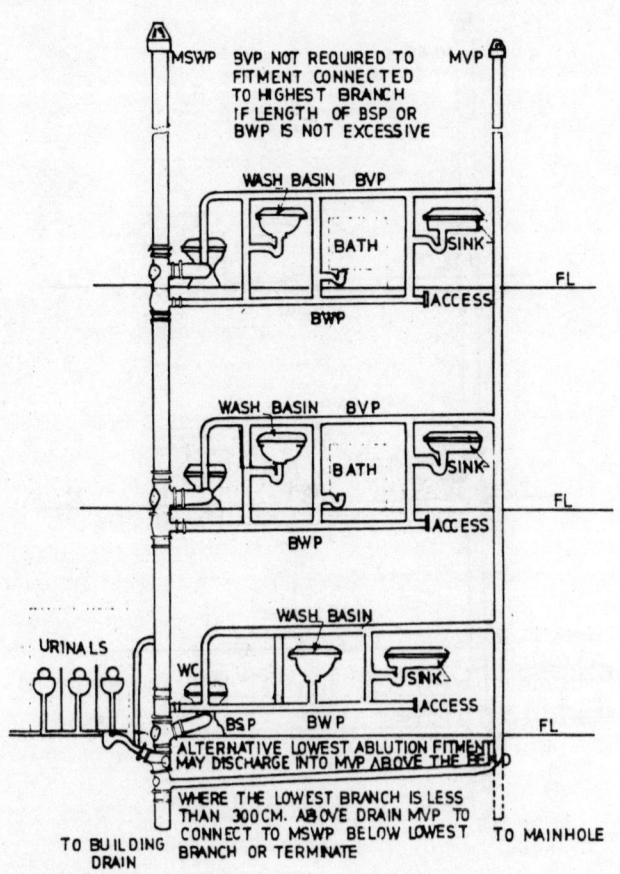

| MWP | main waste pipe | MVP | main vent pipe |
| MSP | main soil pipe | MSW | main soil waste pipe |

Fig. 3.6 *One pipe system.*

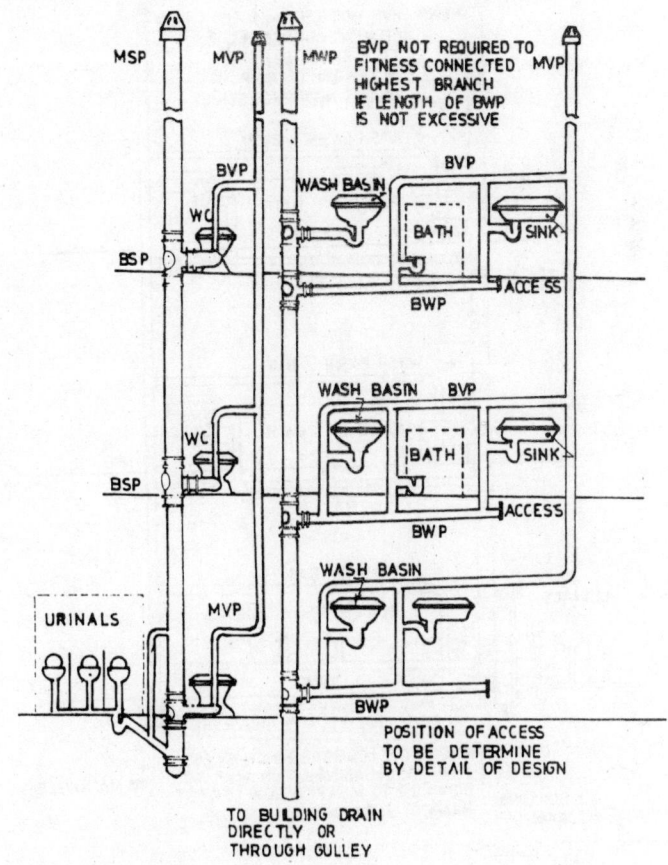

| BVP | branch vent pipe | BWP | branch waste pipe |
| BSP | branch soil pipe | WC | water closet |

Fig. 3.7 *Two pipe System.*

Single Stack System

Single stack system contains a single vertical pipe which acts as main soil cum waste pipe and in addition to it acts as vent pipe. This is a very economical system, but all the traps should have a water seal of not less than 75 mm, to prevent breaking of water seal due to siphonic action (Fig. 3.8B).

Pipelines for soil in buildings should never be less than 4 inches diameter (100 mm). Main waste pipe must be at least 3 inches diameter or 75 mm. Water closets, urinals, bathtubs, washbasins are connected to galvanised iron pipe (preferable) and should be at least 2 inches or 50 mm in diameter. All branch horizontal pipes must have a gradient of at least 1 in 50 and never more than 1 in 10.

The house drain runs at a slow gradient either to a public sewer (most usual in towns) or some form of approved disposal plant. The gradient of the drain should be neither too steep nor too shallow, otherwise blockages and difficulties in disposing of the waste can occur. The optimum "self-cleaning" gradient is normally taken as 1 in 40 for a 100 mm, diameter and 1 in 60 for a 150 mm diameter pipe.

In a town, it is to be connected to the public sewerage system or if this is too far away or too deep, the soil stack is connected to the another form of approved disposal. Alternate forms of sewage disposal are septic tanks and cesspools, neither's a particularly attractive system, especially in towns.

A septic tank is really a miniature chemical treatment plant, and the treated fluid drains away into the ground. It is permitted in the premises, if it would be no way harmful or offensive to other residents in the area.

A cesspool is simply a storage chamber which must be emptied periodically by a special plumbing tanker. So, connecting the waste pipes to a sewer, or a satisfactory alternative, is crucial while planning a bathroom.

Sanitary Fittings

The sanitary fittings indicate all the fittings or appliances used for collection and discharge of soil or waste water. Different sanitary fittings perform different types of functions. They are normally made of ceramics, glazed fire clay, glazed earth ware or glazed chinaware or acrylic. The fittings are

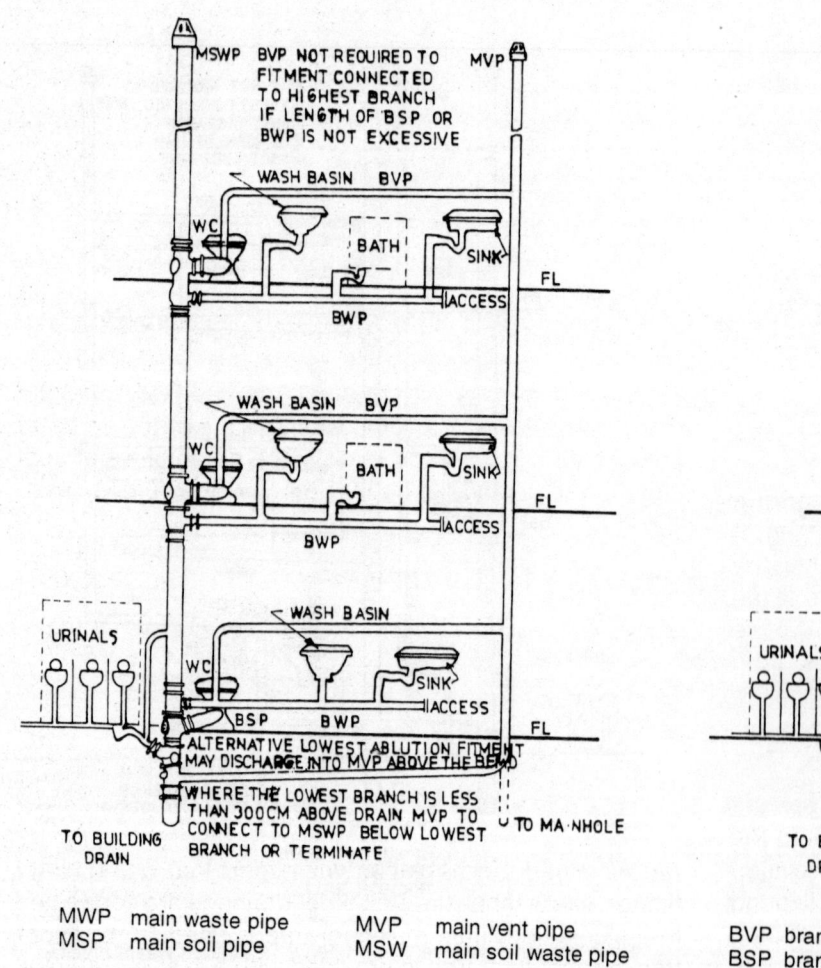

| MWP | main waste pipe | MVP | main vent pipe |
| MSP | main soil pipe | MSW | main soil waste pipe |

Fig. 3.8A *Partially ventilated single stack system.*

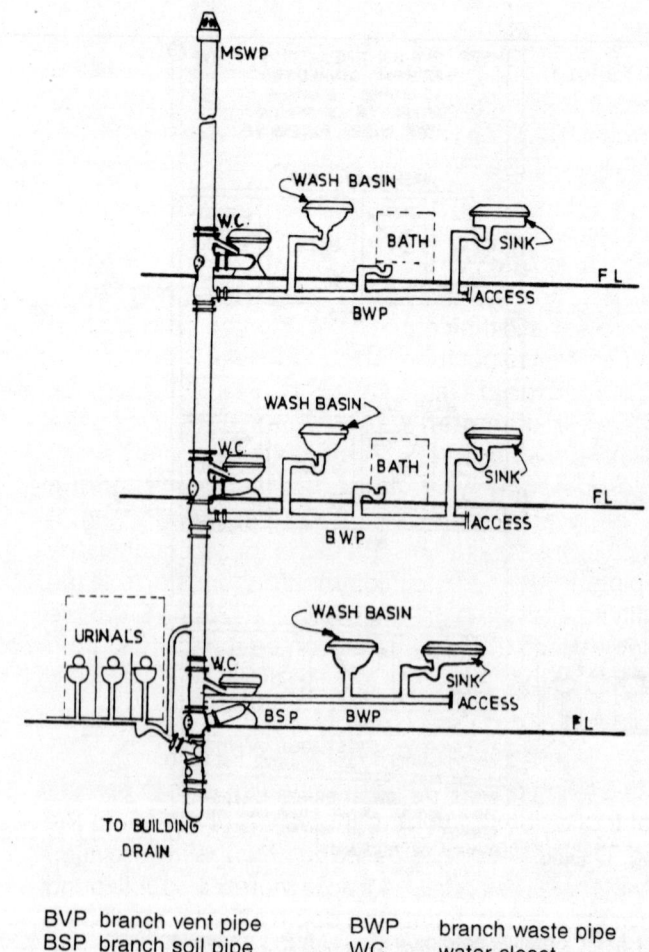

| BVP | branch vent pipe | BWP | branch waste pipe |
| BSP | branch soil pipe | WC | water closet |

Fig. 3.8B Single stack system.

so designed as to have non-absorbent surface which can be easily cleaned. Various sanitary fixtures are:

(i) Washbasin
(i) Waterclosets
(iii) Bathtub
(iv) Urinals
(v) Flushing cistern
(vi) Bidets.

All the fixtures need careful plumbing for their efficiency. These fixtures collect soil or waste and main discharges it to their respective pipes. All the fixtures perform separate functions and have been discussed in details in Chapter 4 on "Sanitary Fittings" separately.

The arrangement of the various fittings is usually decided on the basis of:

(i) How handy the plumbing connections are
(ii) Convenience in use
(iii) General appearance.

But in many cases, it will be the plumbing considerations which will primarily dictate the layout. It is the disposal of waste matter which usually presents more problems than the supply of water. Supply runs are normally in small bore tubing which is fairly flexible as to where you could put it. The WC for example, has a large diameter outlet near the floor and should be positioned as closely as possible to the soil stack it is connected to. The soil stack receives the waste from the WC and carries it to an underground drain. A similar pipe which receives discharge from the washbasin, bath and shower is called a waste stack. Waste stacks have a gully trap at a ground level which incorporates a water seal to isolate the stack from drain gases, whereas soil stacks connect directly to the drain at a manhole.

Bidets also have their outlets near the floor, so it is often most convenient to put them next to the WC. Showers and bathtubs, however, are more flexible in this respect, as they have smaller diameter waste pipes and discharge comparatively cleaner water. Their waste pipe may have to run under floor surface if fitting is at some distance from the stack. The washbasin is most accommodating fitting. It can be placed almost anywhere in the room, because the waste pipe is the smallest in diameter and it starts at the actual fitting and some distance above the floor. From the plumbing aspect, the washbasin can be the farthest from the stack.

Traps

A trap is a fitting provided in a drainage system to prevent entry of foul air or gases from the sewer or drain into the building. The barrier to the passage of foul air is provided by the water seal in the trap. A trap is merely a double bend or loop in the sanitary fitting, the depth of water seal being the distance of the first bend and the bottom of the second.

The deeper the seal the more efficient is the trap. The depth of the water seal vary from 40 to 75 mm. The trap

should always be fitted close to the waste or soil fitting unless the trap is an integral part of the filling as in case of European WC (siphonic type).

Types of Traps

Depending upon the shape. The commonly used traps are P trap, Q trap and S trap, named after the letters they resemble (Fig. 3.9).

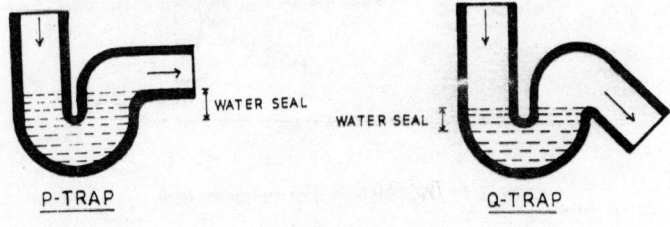

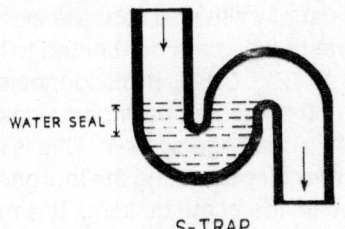

Fig. 3.9 *Different shapes of traps.*

Depending upon the use and location, various types of traps, for the application in bathrooms are as under:

(i) Floor trap (Nahani trap)
(ii) Multifloor trap
(iii) Gulley trap.

Floor trap   These are provided in floors to collect waste water from kitchen sinks, bathroom floors, washing floor, etc. A floor trap forms the starting point of waste water flow. The Nahani trap is made of cast iron or polyvinyl chloride (PVC) provided with a removable grating at top so as to prevent the entry of solid matter. The depth of water seal of floor trap should not be less than 40 mm (Fig. 3.10).

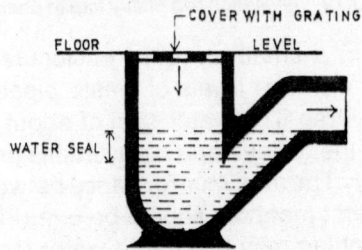

Fig. 3.10 *Details of floor trap.*

Multifloor trap   In the places where more than one sanitary fittings, like washbasin, sink, floor drainage are connected to a single floor trap, multifloor trap is used. In multifloor trap, normal floor trap is provided with two or more holes to be connected to the waste pipes below the floor level Use of multifloor trap ensures leakproof joint (Fig. 3.11).

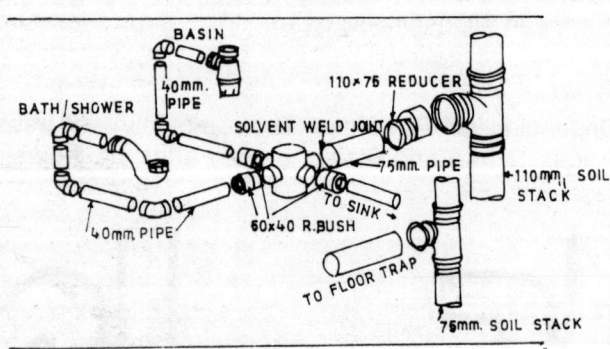

Fig. 3.11 *Typical layout of multifloor trap.*

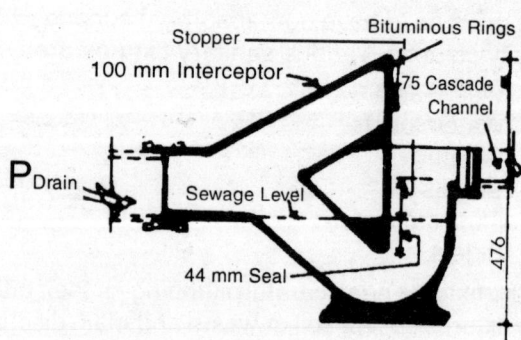

Fig. 3.13 *Intercepting trap.*

Gulley trap It is usually situated near the external face of the wall. All the waste pipes are connected to house drain through gulley trap (Fig. 3.12). It disconnects the waste water flowing from the kitchen, bathroom, washbasin and floors from the main drainage system. This is a deep seal trap forming a barrier for preventing the foul gases from the house drain to the inside of the building. It is made of cast iron or glazed stoneware. Grating is provided on top to retain all solid matter.

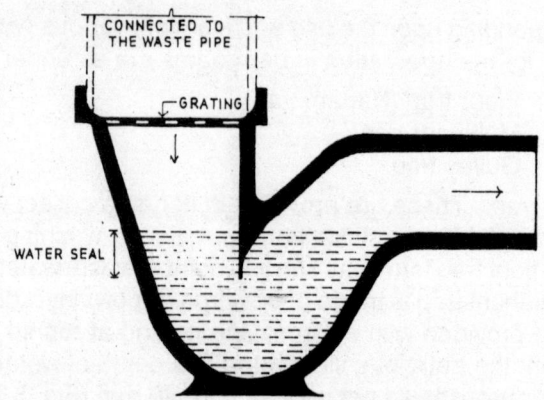

Fig. 3.12 *Installation of a gulley trap in drainage.*

It is fitted in a small masonry enclosure to meet the requirements of invert levels of waste pipes discharging into the gulley trap. The water seal of about 60 to 70 mm is provided in the gulley trap. Gulley trap is provided in the waste pipe only. The maximum distance between the gulley trap and the first manhole should be 6 m (Fig. 3.13).

Faulty plumbing may result into water leakages in the walls and floors of the bathroom leading to the dampness in adjoining rooms and lower floors as well. Moreover, it may result into cumbersome maintaining problems. Thus, extra care is needed at the time of installation of plumbing system for an efficient functioning of the bathroom.

- The pipelines should be as straight and short as possible and that all joints be absolutely watertight for the efficient working of the plumbing system.

- The water supply pipe and drainage pipes should not be in close vicinity of each other so that in case of leakage, water does not get polluted.

- Extra precautions must be taken against leakages, especially where pipes are concealed. The best way is to paint all GI pipes with anticorrosive paint. The threaded ends must be given two coats before joining and another coat at the time of joining with the socket.

- The threaded ends of pipes and stopcocks are sealed with zinc paste have coated hessian strings. The strings are been replaced by teflon tape or polytetrafluoroethylene (PTFE) tape, which is more durable and is a better seal.

- Before sealing grouted pipes into walls, it is necessary to test the pipes by pushing water in the pipes under great pressure. The pressure test will show any minute holes that the grout, the coating or the naked eye might have missed.

- If you have some exposed pipes in the bathroom, try not to let them run for more than 6 feet area to hide the pipe and also function as a bathseat or a foot rest.

- The pipe connected to the bibcock should not be less than half a inch or 12 mm in diameter. All taps should be fitted on the corners of the tiles so that, while fitting, tiles do not break, this also makes for neatness.

- For the rising main of the overhead tank one and a half inch or 40 mm GI pipes are used, and for the overflow, outlet and washout pipes, one inch or 25 mm GI pipes are used. Pipes bring water from the overhead tanks to the bathroom and are systematically reduced from one and a half inch to half inch diameter by reducers.

- A squatting pan should be kept as close to the external wall as possible.

- A service pipe should never lead directly to a water closet, for which water must come through a flushing cistern. All flushing cisterns must be supplied from overhead tanks through down take pipes.

- After a water closet is installed, it is filled with water to its normal water seal level. Pieces of tissue paper or toilet paper are crumpled into a ball of 6 or 150 mm diameter along with some cork bungs are put in the toilet bowl and flushed. If the flush carries away all this material at least 75 per cent of the time, the functioning of the system is considered satisfactory.

- The distance of the cistern for washing down a pan should at least 6 or 1800 mm from the top of the pan to the bottom of the cistern. For the siphonic type of water closet, a low level cistern and commode should be about 1 or 300 mm. Flushing pipes should not be less than 32 mm in diameter for the wash down system and at least 40 mm for the low level siphonic type.

- Almost all supply pipes enter the bathroom at one point and then branch out to the various points where the sink, bath area or tub, or water closets are to be installed. Ideally, pipes should run along the walls, instead of under the flooring to maintain sufficient water pressure in the taps.

- Use of storage geysers are very common for hot water supply. While connecting geyser to piping, there should always be no-return valves on both inlet and outlet pipes to avoid hot water running into empty cold water pipes and vice versa.

- Pipe lines in soil in building should never be less than 4 inches in diameter (100 mm). The main waste pipe must be at least 3 inches in diameter or 75 mm. Water closets, urinals, bathtubs, washbasins, are connected to galvanised iron pipe (preferable) and should be at least 2 inches or 50 mm in diameter.

- The overflow pipe of the bathtubs should be connected properly to the floor trap or the waste pipe. No pipe should be left open around bathtubs.

- The main source of leakage in the bathrooms is the floor trap. Thus, it is necessary that all the connections are properly sealed with flexible waterproofing materials to avoid the water leakage.

- Irrespective of the material used, all water and drainage pipes should be tested by hydraulic test, before providing wall and floor treatment to the bathroom.

SANITARY FITTINGS

The various sanitary fittings which are used in the bathrooms are discussed in this chapter. The various type of bathroom fittings depending upon its application on the basis of space requirements, economics and utility are explained. The various bathrooms fittings are as under:

- Washbasins
- Water closets
- Urinals
- Bathtubs
- Bidets
- Flushing cisterns

WASHBASINS

A washbasin is an appliance used for washing hands, face, etc. It is available in wide range of patterns and sizes. Washbasin can be either made of glazed chinaware, ceramic enamelled cast iron or stainless steel. The most common type of washbasin is wall mounted which is simply mounted on angle irons fixed on the wall. The wall-mounted washbasin can be connected to the floor trap by either a flexible waste pipe which is directly inserted through the hole of the grating of the floor trap or by concealed waste pipe through bottle trap. The bottle trap is exposed, thus, sometimes not very appealing aesthetically. There is another kind of washbasin which is supported on the pedestal. The function of the pedestal is to conceal the bottle trap. The washbasins further modifies into counter types which apart from concealing the bottle trap, can be used for keeping day to day toiletries (Colour Plate 1).

The height of the washbasin depends on the height of user and is usually kept 750 to 800 mm above the floor level.

The washbasins are available with single hole, or three hole for the tap connections for hot and cold water supply. Sometimes, washbasins are available with semipunched holes as well, so that the holes can be punched at site to suit the type of the faucet.

Wall-mounted Washbasin

Wall-mounted washbasins are either supported on CI brackets with stud from wall or rag bolts grouted in the wall

or the washbasin is directly screwed to the wall. The waste pipe from the washbasin is connected to the floor trap either directly or through a bottle trap. Figures 4.1A, B and Fig. 4.2 show the alternative details.

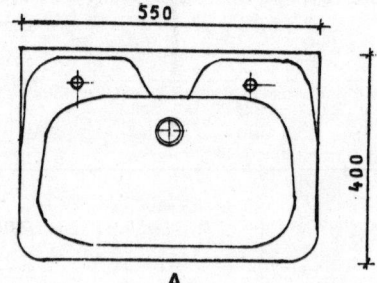

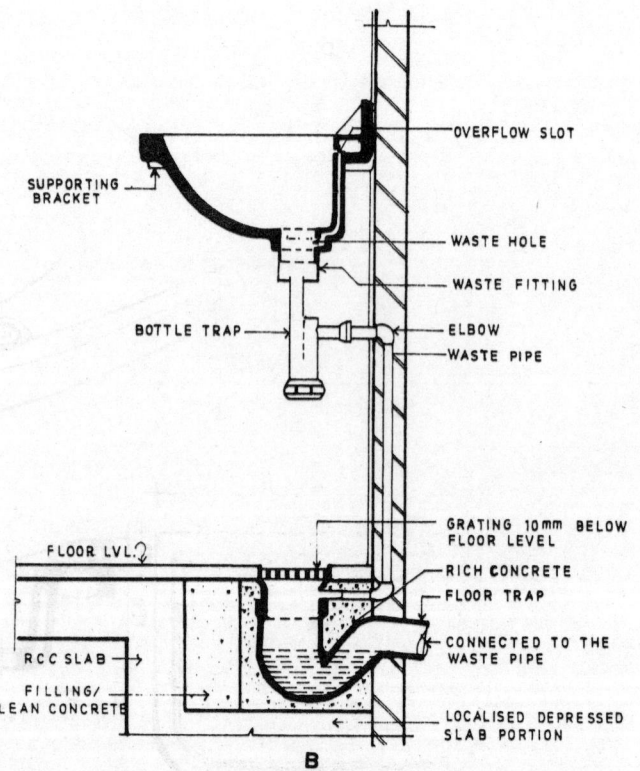

Fig. 4.1 *Installation of washbasin showing various connections.* **A** *Plan* **B** *Section showing waste pipe concealed from view.*

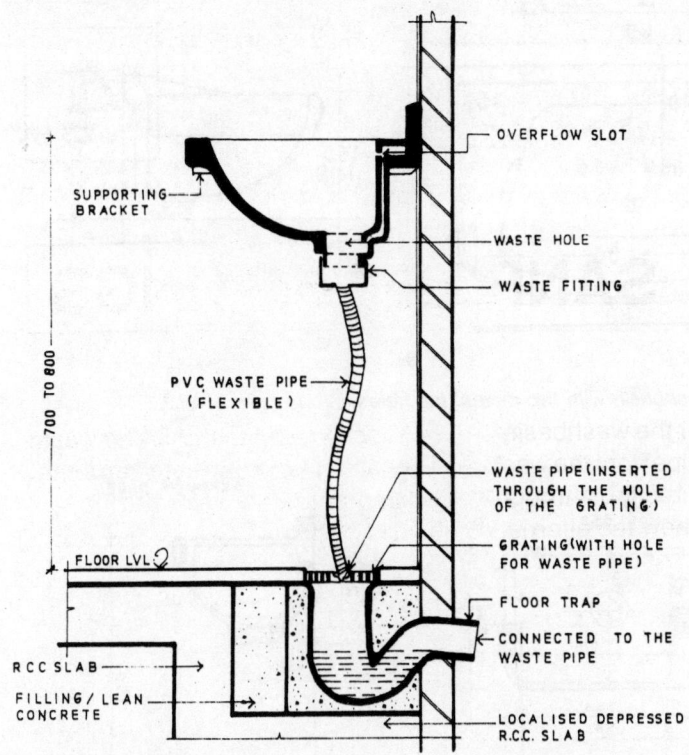

Fig. 4.2 *Typical vertical section of washbasin (waste pipe open to view).*

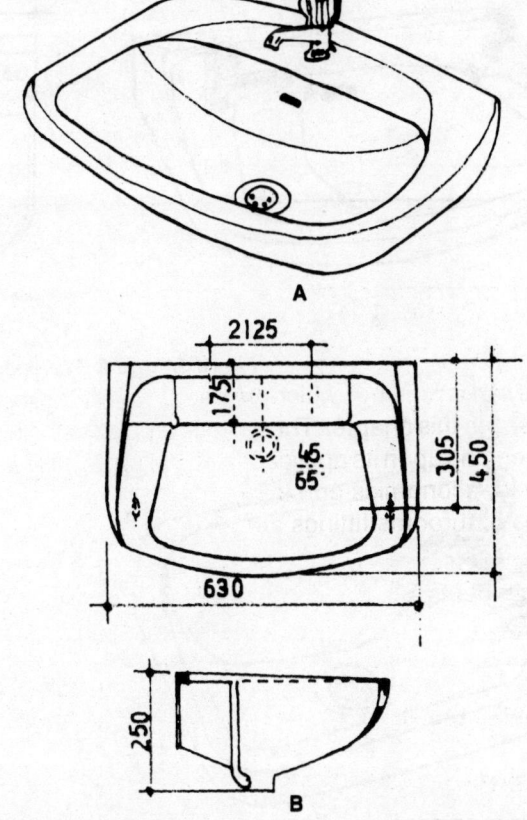

Figs 4.3A and B *Wall mounted washbasin with centre tap hole.*

A variety of washbasins are available in the market under different trade names. Washbasins are of vitreous china available with punched or semipunched holes for the fixing of taps. Number of holes depends on the type of fittings. Following are the few basic types of washbasins (Figs 4.3A to 4.7B).

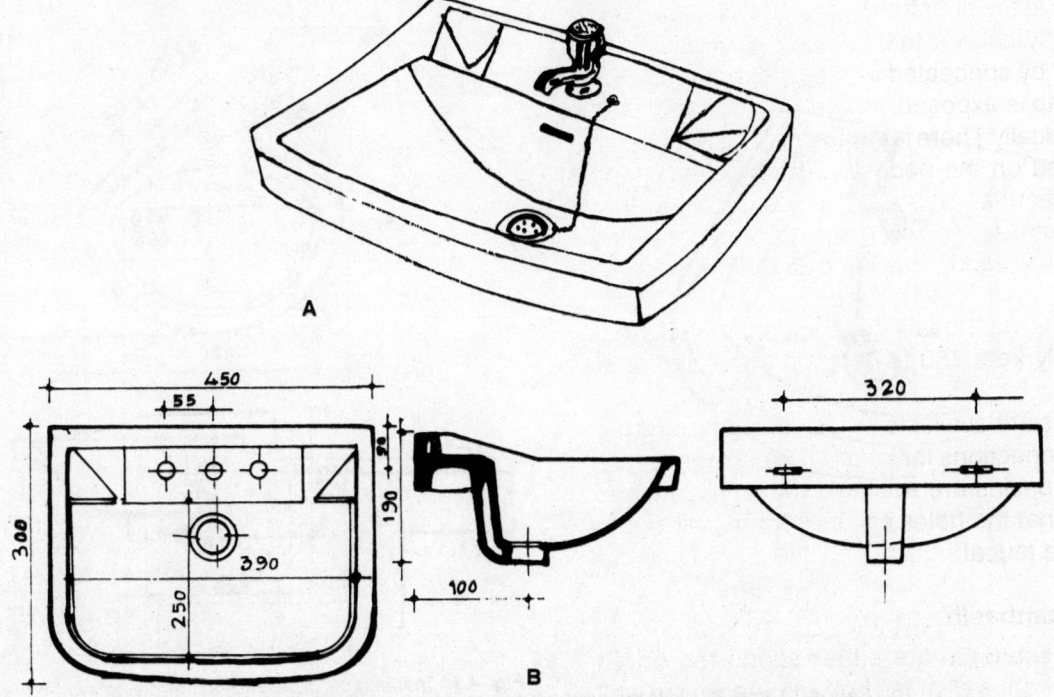

Figs 4.4A and B *Wall mounted washbasin with three semipunched holes to suit various fittings.*

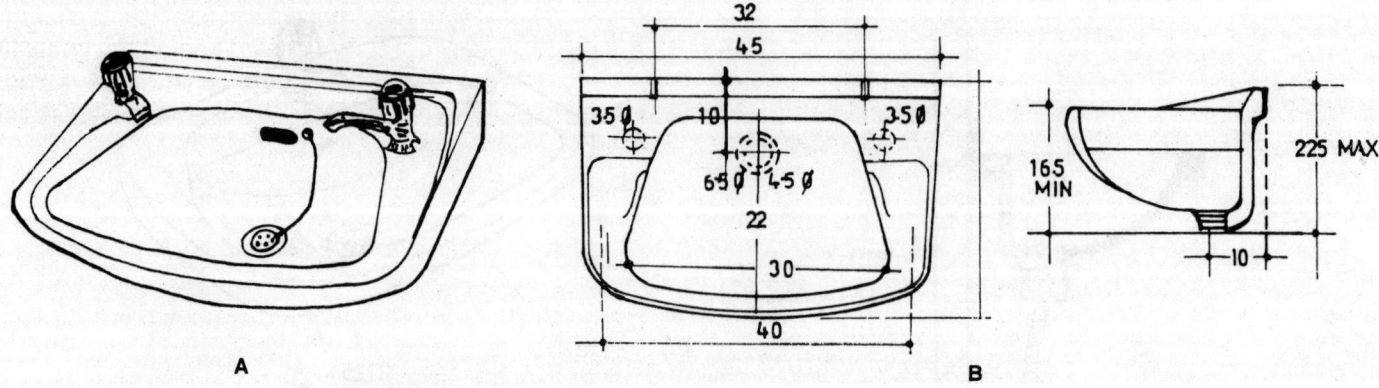

Figs 4.5A and B *Wall-mounted washbasin with two distant tap holes.*

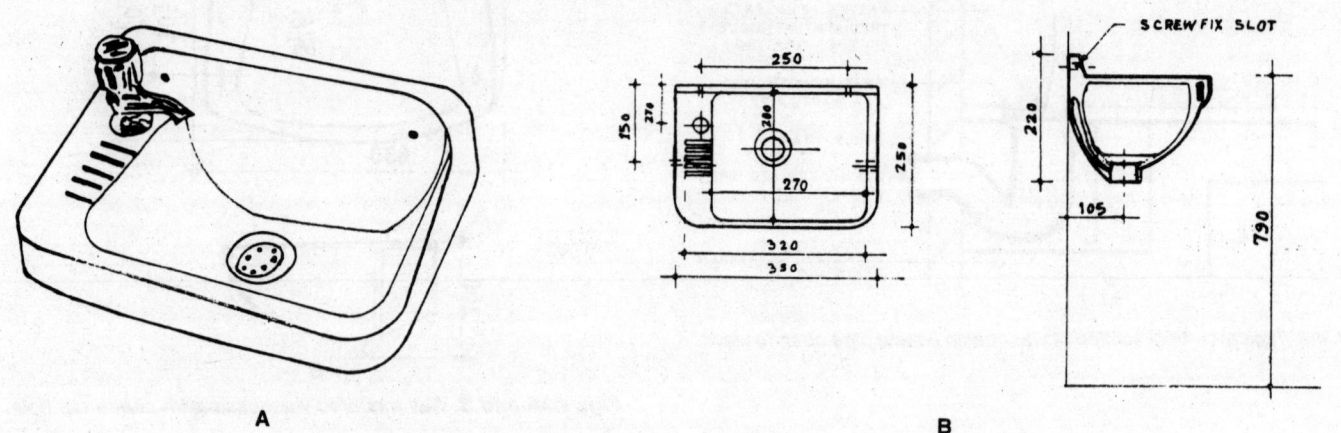

Figs 4.6A and B *Wall-mounted washbasin with one-side tap hole.*

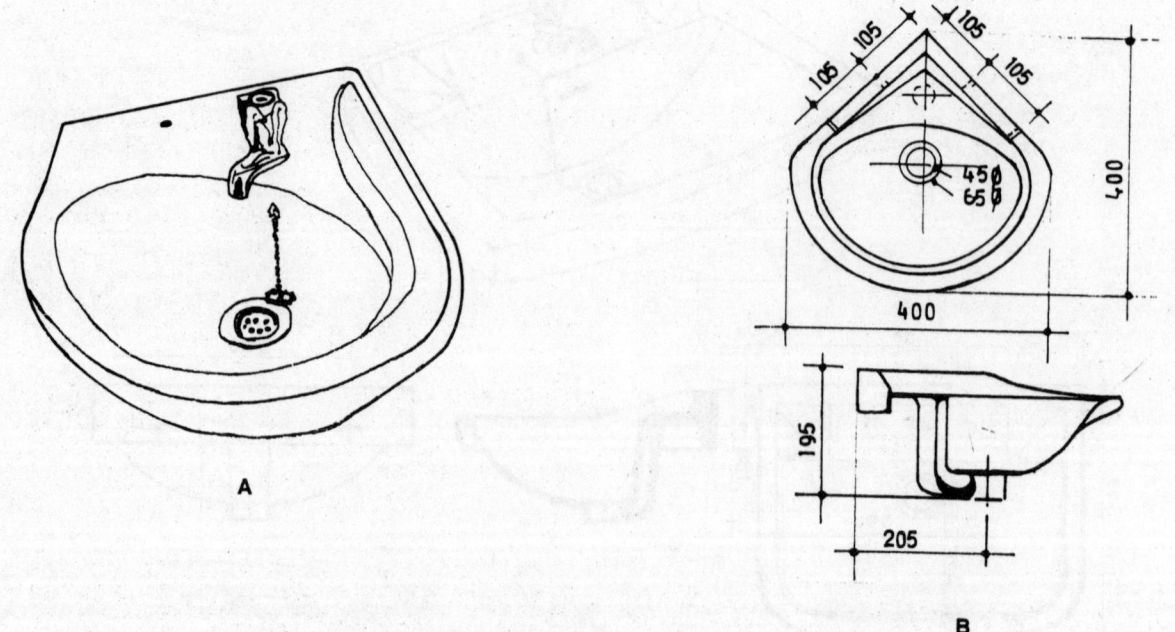

Figs 4.7A and B *Wall-mounted corner washbasin with single tap hole.*

Normally, washbasins are available in three ranges: small, medium and large. Table 4.1 depicts the common dimensions of the small, medium and large washbasins in mm.

The advantage of pedestal type washbasin is that the wastepipe or bottle trap is concealed behind the pedestal and gives a very neat look. But the slope of the flooring around the pedestal should be perfect so that water does not collect in front of the pedestal.

Table 4.1: *Common dimensions of washbasins (in mm)*

	Small	Medium	Large
A	400–450	500–550	600–650
B	300	400	400–500
C	200	225	250

Pedestal Type Washbasin

Pedestal type washbasin is also supported on the rag bolts or CI brackets (Figs 4.8A to 4.12B). This type of washbasin is generally available in medium and large sizes (Colour Plate 2).

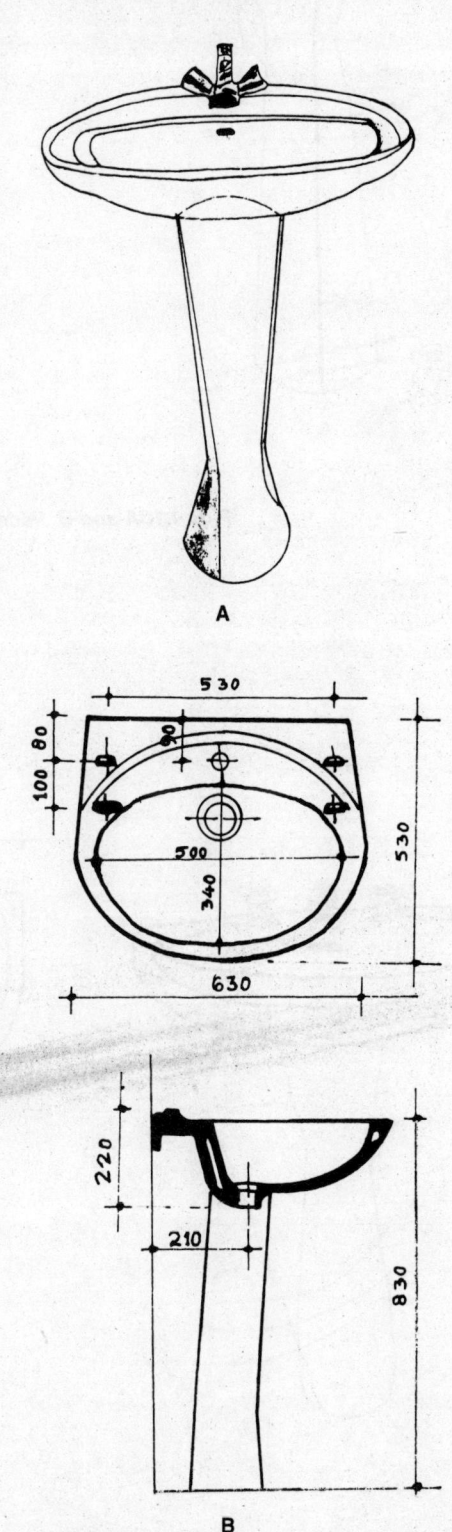

A

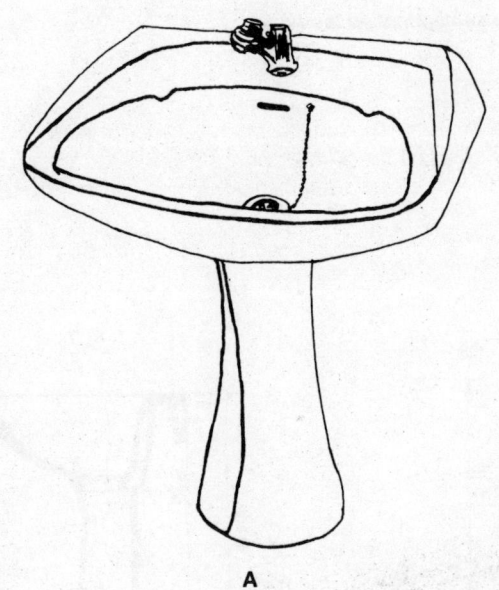

A

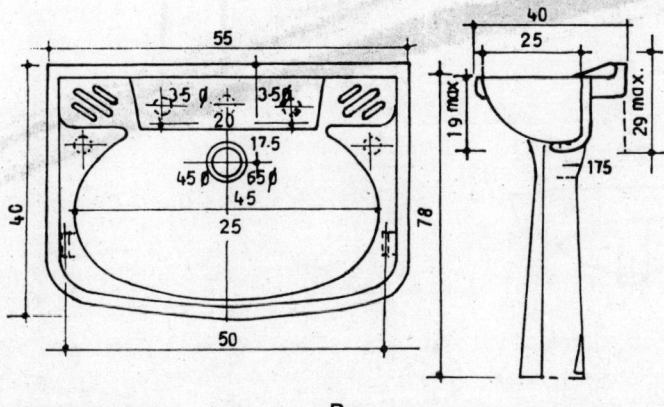

B

Figs 4.8A and B *Pedestal washbasin with centre tap hole and four semipunched holes.*

Figs 4.9A and B *Pedestal washbasin with centre tap hole.*

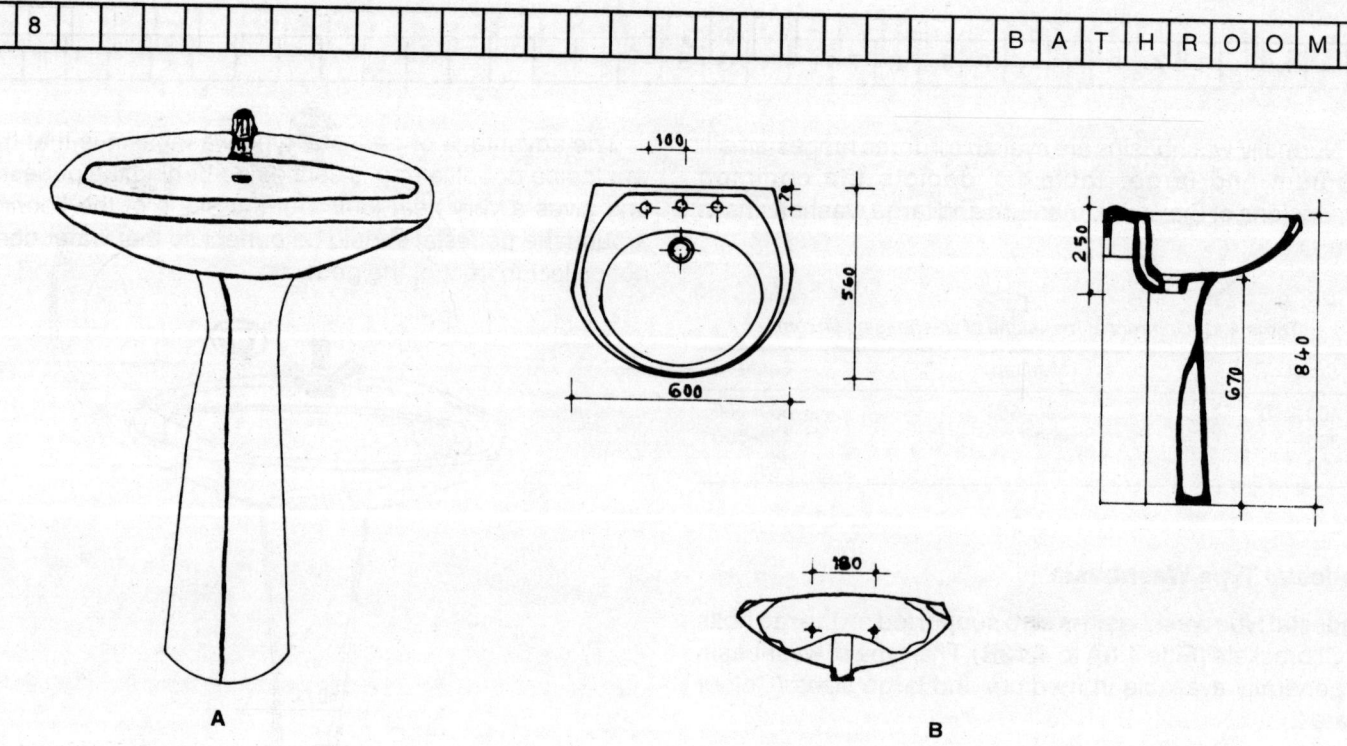

Figs 4.10A and B *Pedestal washbasin with three semi-punched tap holes.*

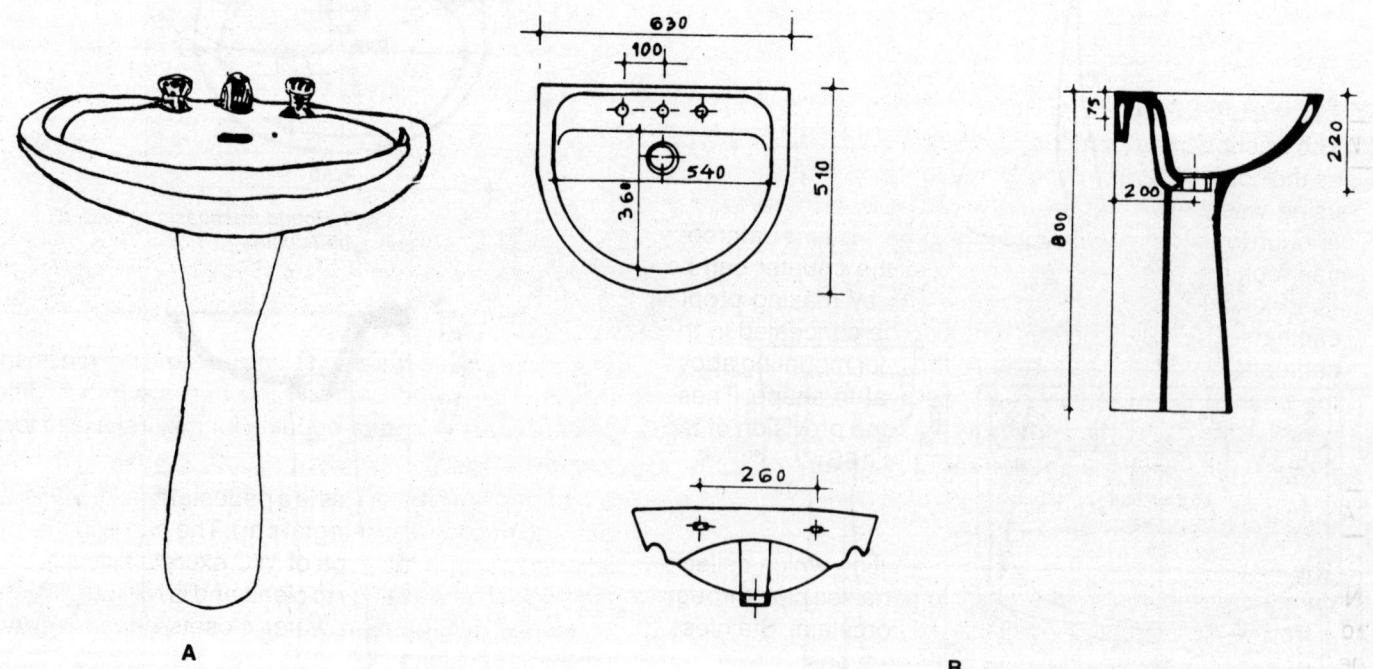

Figs 4.11A and B *Pedestal washbasin with three tap holes.*

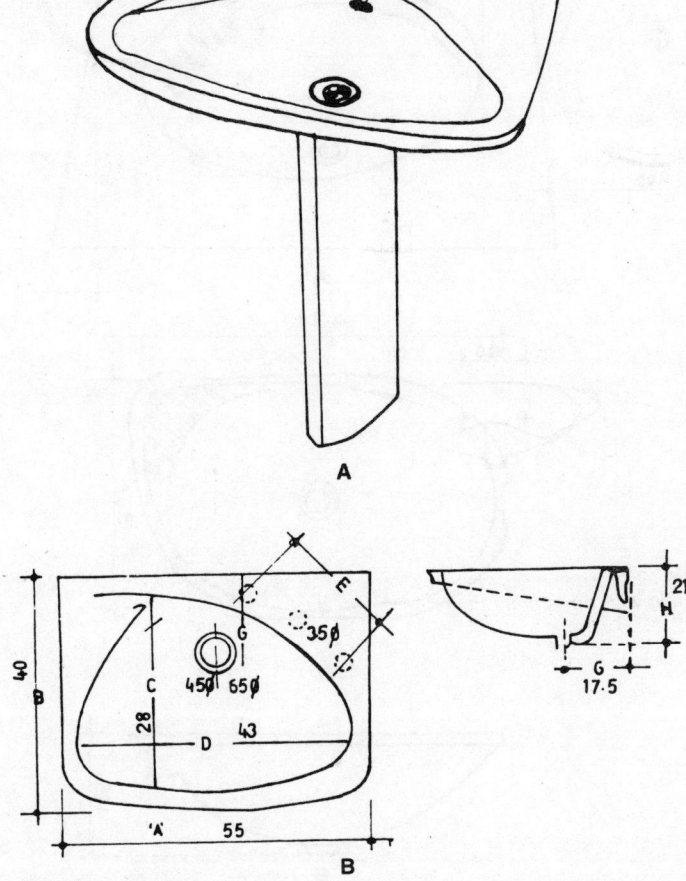

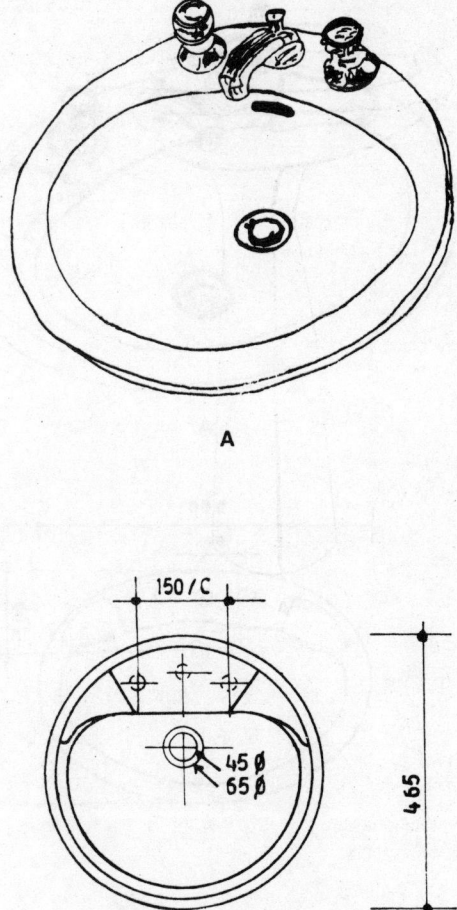

Figs 4.12A and B *Pedestal washbasin with three semi-punched side holes.*

Figs 4.13A and B *Round washbasin suitable for mounting above the counter.*

Counter Washbasins

The washbasin with a counter is better than the above type as the counter can be used for the things you require while using washbasin. But the bathroom should be spacious enough to accommodate a counter otherwise the bathroom can look cramped. The space under the counter can be used for the storage of toiletries, etc. by making proper cabinets. The waste fittings can also be concealed in the cabinets. Washbasins that are suitable for mounting above the counter are normally round or oval in shape. These washbasins are either with or without the provision of tap holes in the basin rim (Figs 4.13A to 4.16B).

WATER CLOSETS

The water closet (WC) is a sanitary fitting which collects human excreta and discharges it into the soil pipe through a trap. Water closet is available in porcelain, Stainless, Steel, etc. The water closet is of three types:

(i) Indian type water closets
(ii) European type water closets
(iii) Anglo-Indian water closets

Indian Type WC It is a squatting pan and is fixed flush with the floor level. The pan and the trap are in two different pieces. The trap has an opening for making a joint for anti-siphonic pipe.

European Type WC This is a pedestal type of pan in which pan and trap form an integral part. The pan is in the form of inverted cone. In this type of WC excreta falls directly into the trap and thus is easy to clean and is hygienic. There are two types of European water closets depending on the system of flushing.

(i) Wash down: In this type of water closets, the pan are removed by the gravity flush of water. This type of water closets makes a lot of noise while flushing.

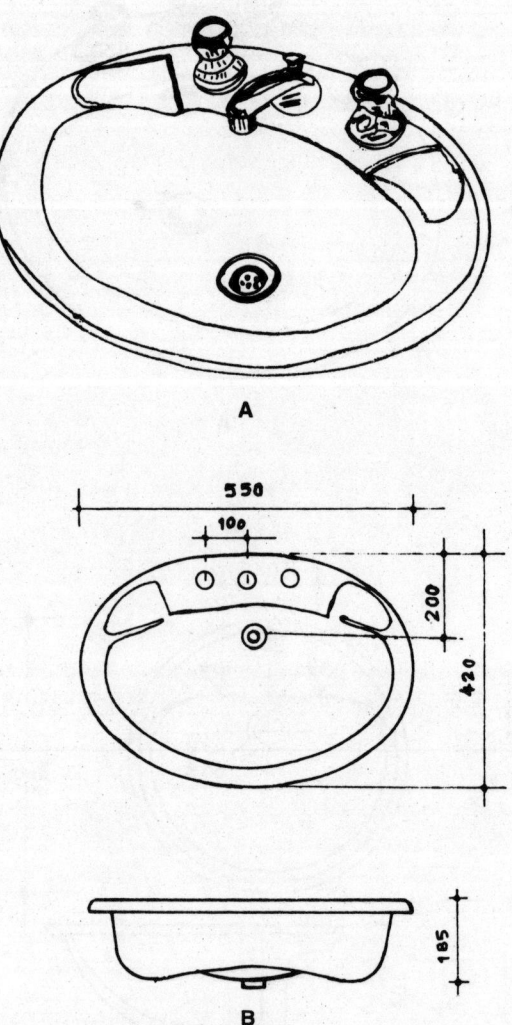

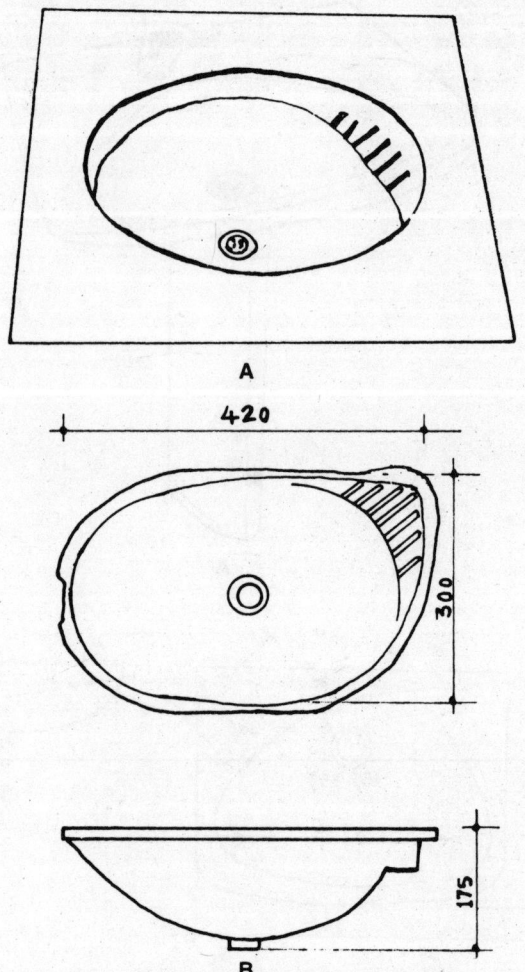

Figs 4.14A and B *Oval washbasin suitable for mounting above the counter.*

Figs 4.15A and B *The above oval washbasins are without tap holes. Tap holes to be provided in the counter.*

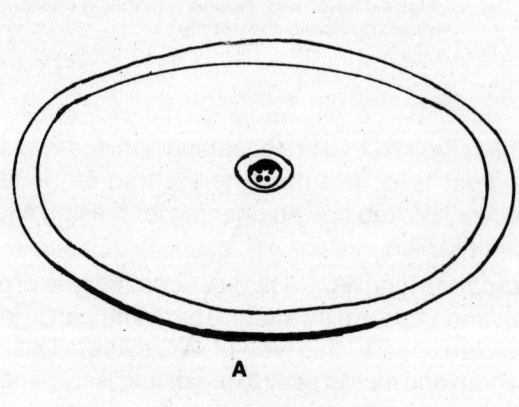

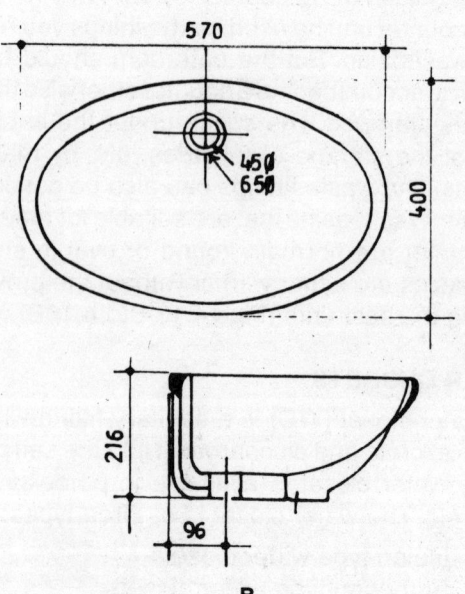

Figs 4.16A and B *Holes are to be provided in the elliptical counter.*

(ii) Siphonic type water closets: It can be either single siphonic or double siphonic depending on the number of traps used. In this type the contents of the pan are removed by siphonic action when the water closets is flushed and water passes through the pan. Single siphonic type of water closet makes less noise at the time of flushing whereas the double siphonic type gives a silent operation.

After flush chamber in the fitting is provided to re-seal the trap.

The water closets either connected to the flushing cisterns by an exposed flush pipe or are available with close coupled cisterns. Water closets with close coupled cisterns are aesthetically appealing and give a noise-free flushing operation (Colour Plate 3).

Anglo-Indian WC: It is a pedestal fitting with a built-in trap and can be used in squatting as well as sitting position.

It serves the dual purposes, as it can be used as Indian as well as European type of water closet.

The water closets are available in the market in a wide range of designs and colours to suit the decor of the bathrooms.

Indian Type WC

Indian type WC is usually made of porcelain. The WC pan is fixed flush with the flooring and the top of the pan has a flushing rim having number of holes to spread the flush water. The excreta do not fall directly into the trap and there are chances for excreta to become foul, if properly not flushed. The contents of the pan are removed by gravity flush of water. It is fixed in squatting position at floor level. A pair of foot rests are provided on either side of the pan for convenience. Indian WC is available in many types (Figs 4.17A to 4.21B). (*See* Colour Plate 4).

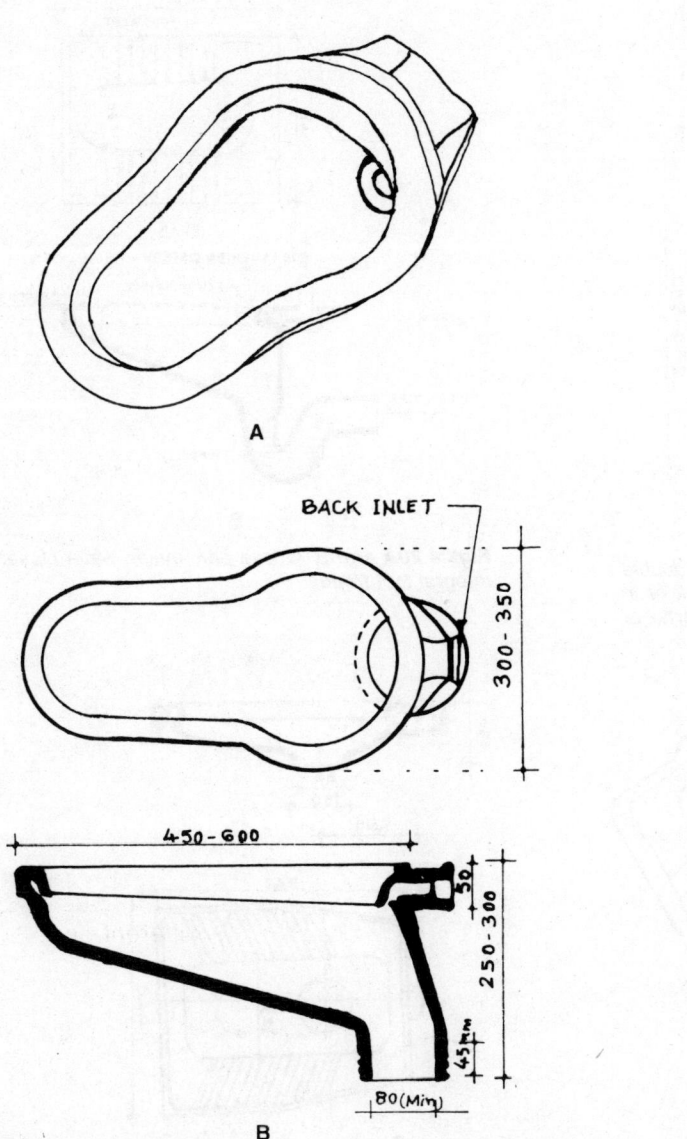

Figs 4.17A and B *Madura pan: Indian water closet with back inlet.*

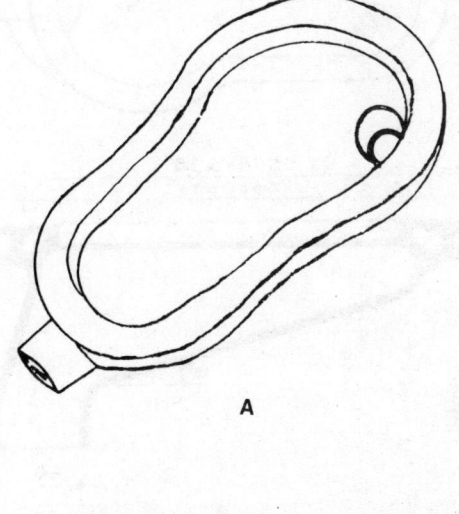

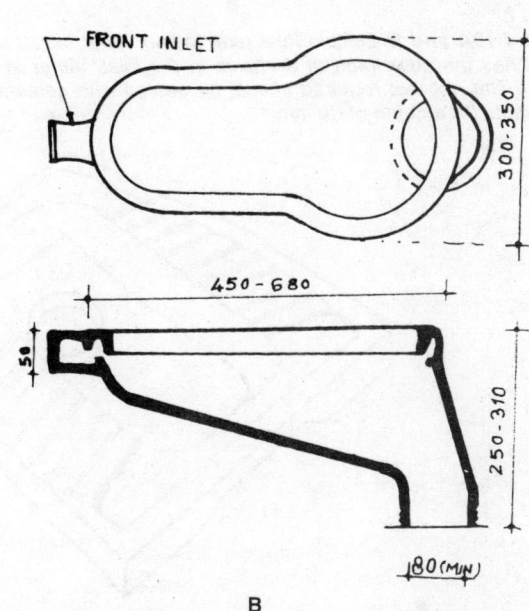

Figs 4.18A and B *Bombay pan: Indian water closet with front inlet.*

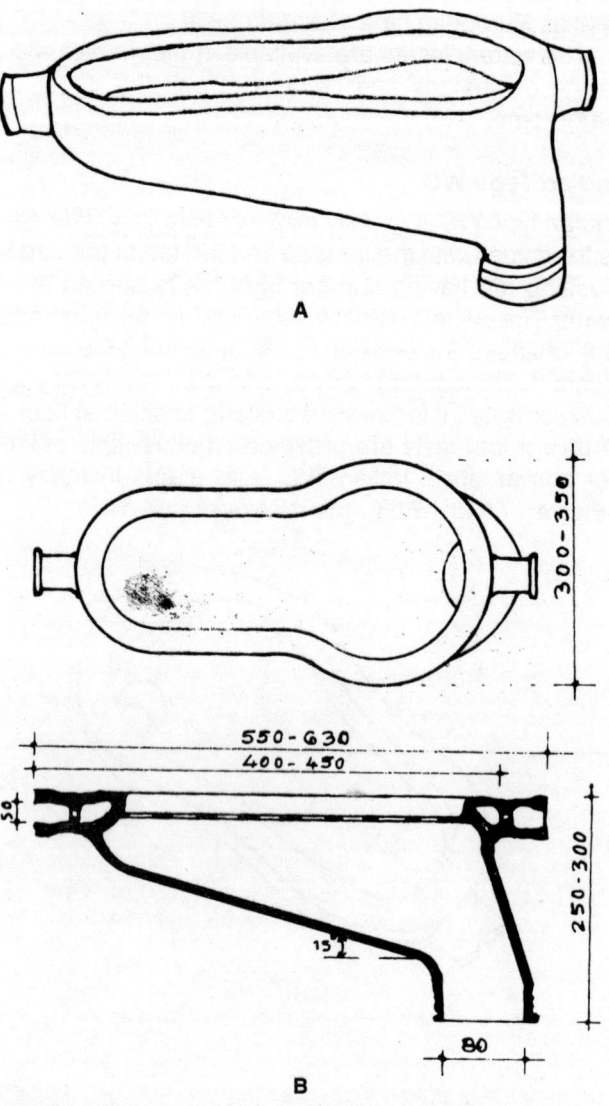

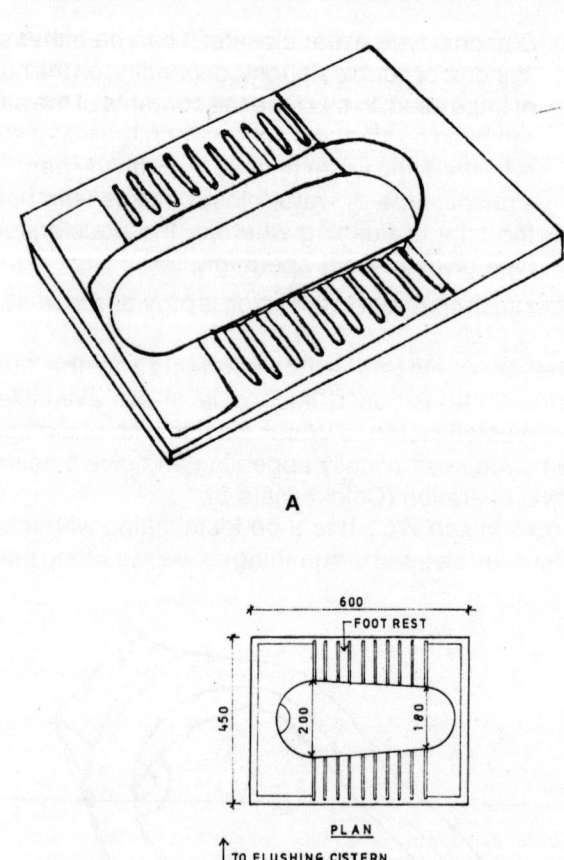

Figs 4.19A and B *Double inlet pan: Indian water closet with double inlet has the advantage of a choice in the inlet, either at back or in front. The one not required should be sealed with cement mortar or sealants for a depth of 30 mm.*

Figs 4.20A and B *Orissa pan: Indian water closet integral foot treads.*

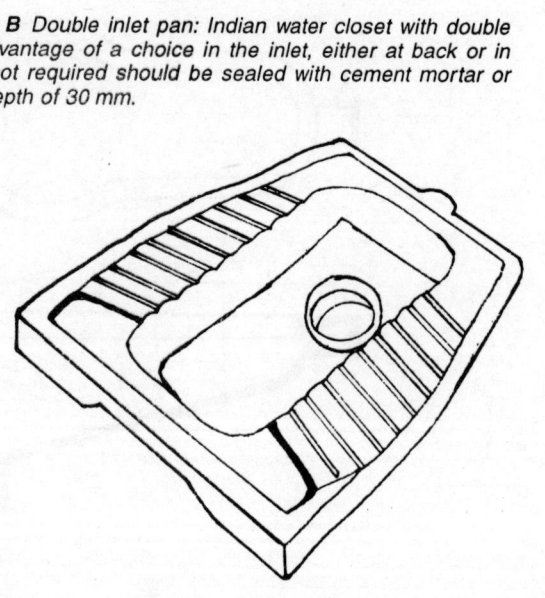

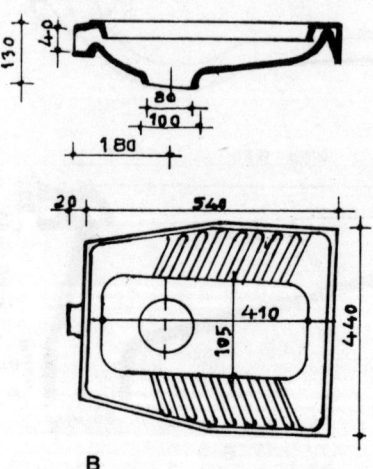

Figs 4.21A and B *Asian pan: Indian water closet suitable for multi-stored buildings because of its low over all height.*

European Type WC

The top of the pan has a flushing rim to spread the flush water. European WC is used in a sitting position over a plastic seat hinged to the fitting. Now-a-days, European WC with ultra-low flush technique is in the market as a water conserving system. It is a water saving water closet and cistern, designed in such a way that it uses only 5 litres of water instead of 10 to 15 litres for flushing the contents of the pan.

There are two types of European WC depending on the way, the contents have been removed from the pan.

Wash Down Type

For ground floor WC fitted with S-trap is used, whereas, for upper floors WC fitted with P-trap is used. The wash down type European water closet is of different types as shown in (Figs 4.22A to 4.24B).

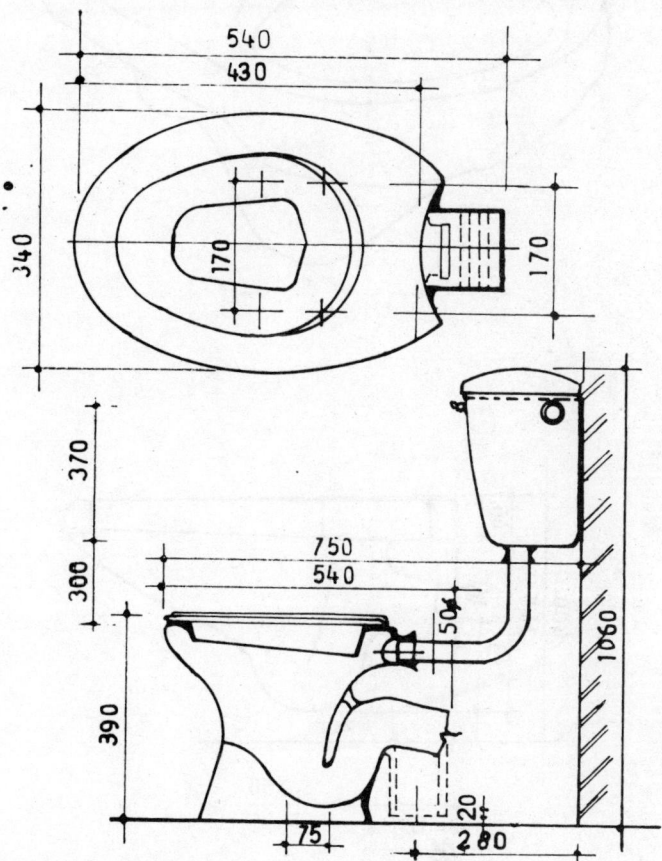

Fig. 4.22 *Wash down type: European water closet is available in 'P' and 'S' trap. To be attached with the cistern with a flush pipe. This type of closet can be used both with low level cistern and high level cistern.*

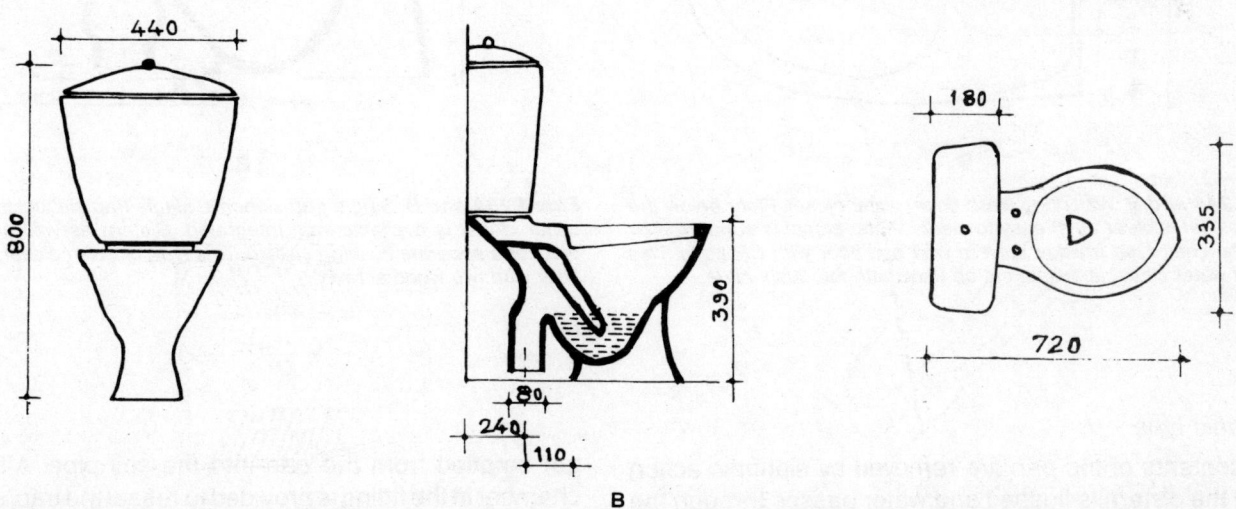

A

B

Figs 4.23A and B *Wash down water closet with close couple cistern. In this type of water closet, flushing cistern is integrated with the water closet with the bottom inlet. This type of water closet is aesthetically better as the flushing pipe is not visible.*

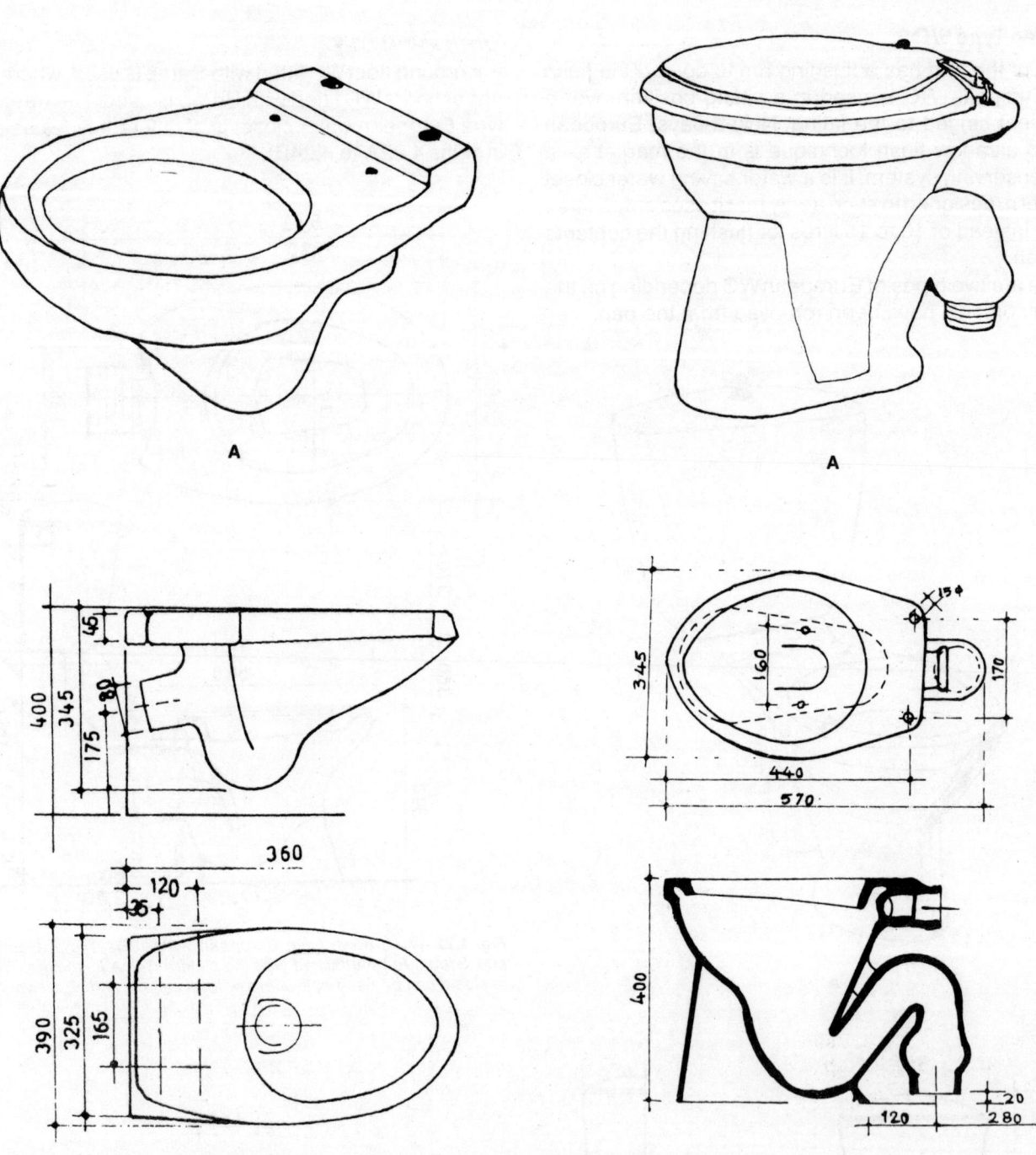

Figs 4.24A and B *Wall hung wash down water closet. Floor below the water closet is clear thus, easy to clean. Water closet is supported by cast iron chair type bracket sunk in wall and floor with CP bolts. This type of water closet is suitable to be used with the flush value.*

Figs 4.25A and B *Single trap siphonic: single trap siphonic European water closet is available with integrated. Cistern as well as flushing pipe for a separate flushing cistern. This type of water closet is fixed to floor with two wood screws.*

Siphonic Type

The contents of the pan are removed by siphonic action when the cistern is flushed and water passes through the pan. The specially built trap sets up siphonic action when water is flushed and the entire water along with the contents get emptied from the pan into the soil pipe. After flush chamber in the fitting is provided to reseal the trap, siphonic type WC may have single and double trap (Figs 4.25A to 4.26B).

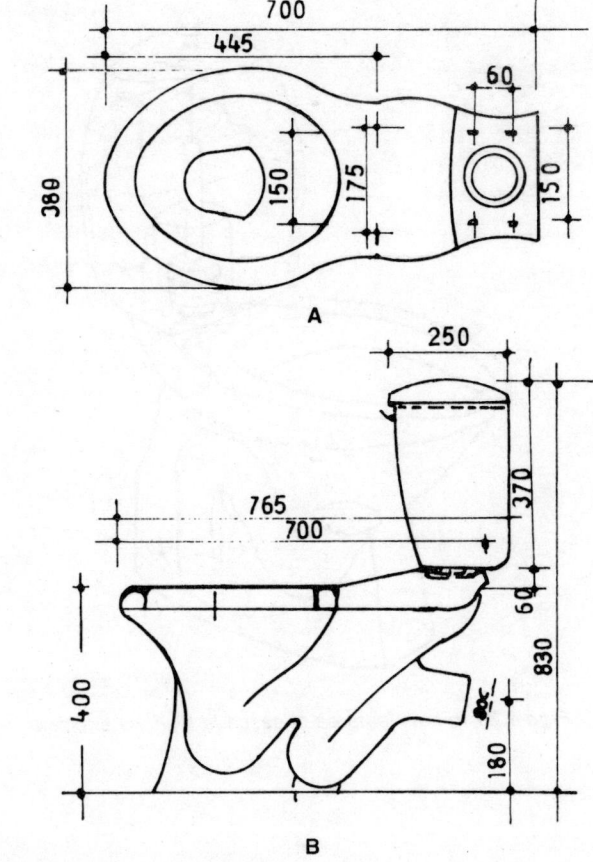

Figs 4.26A and B *Double trap siphonic: This is the double trap siphonic European water closet, close coupled with the cistern. It is fixed to the floor with four wood screws.*

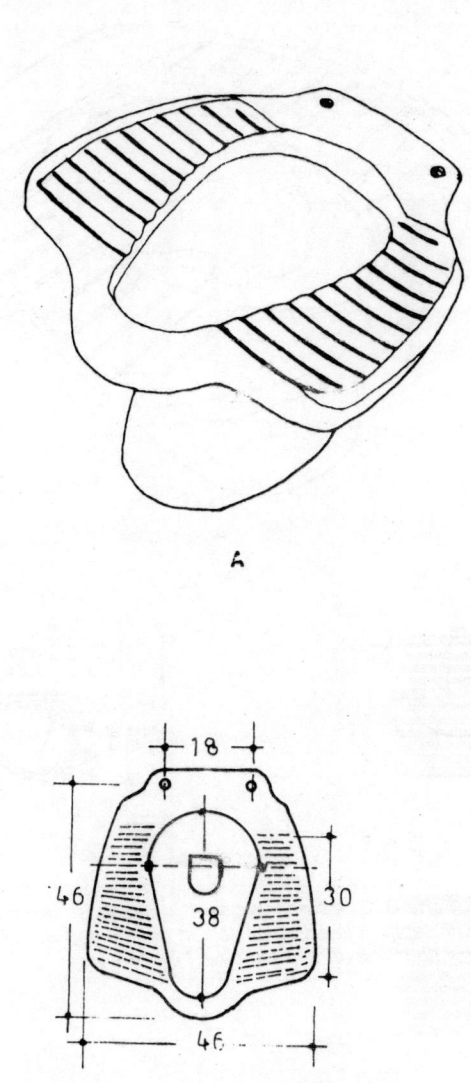

Anglo-Indian Type WC

The top of the pan is provided with a pair of foot rests, whenever it is needed, the plastic seat hinged to the closet can be placed on the foot rests to use it as a EuropeanWC. The excreta falls directly in the trap and thus can be easily flushed out (Figs 4.27A to 4.28B).

Cascade (Ultra low flush)

Cascade is a water conserving system, the elliptical cistern is designed to operate with just 5 litres of water instead of 10 to 12 litres in the conventional cistern. At an average of 20 flushes a day, a single cascade cistern can save 100 litres of water per day for the household. In large establishments like hotels and hospitals, the significance of water conservation is even greater.

The cistern is operated by lifting a knob, water is then propelled through a bifurcated pipe which eventually flushes the bowl. And the waste is expelled through a large circumference trap for total cleansing (Figs 4.29 to 4.32).

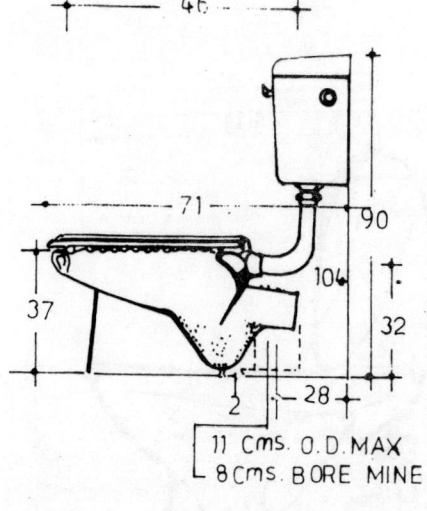

Figs 4.27A and B *Anglo India type: It is a pedestal fitting with in built trap and can be used in squatting as well as sitting position. The top of the pan is provided with a pair of foot rests: Wherever it is needed, the plastic seat cover, hinged to the closet can be placed on the foot rests to use it as a European water closeter. The excreta falls directly in the trap and thus can be easily flushed out and functions like a European*

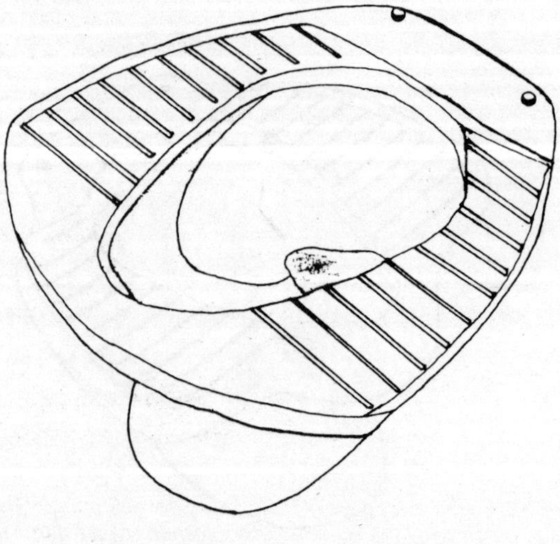

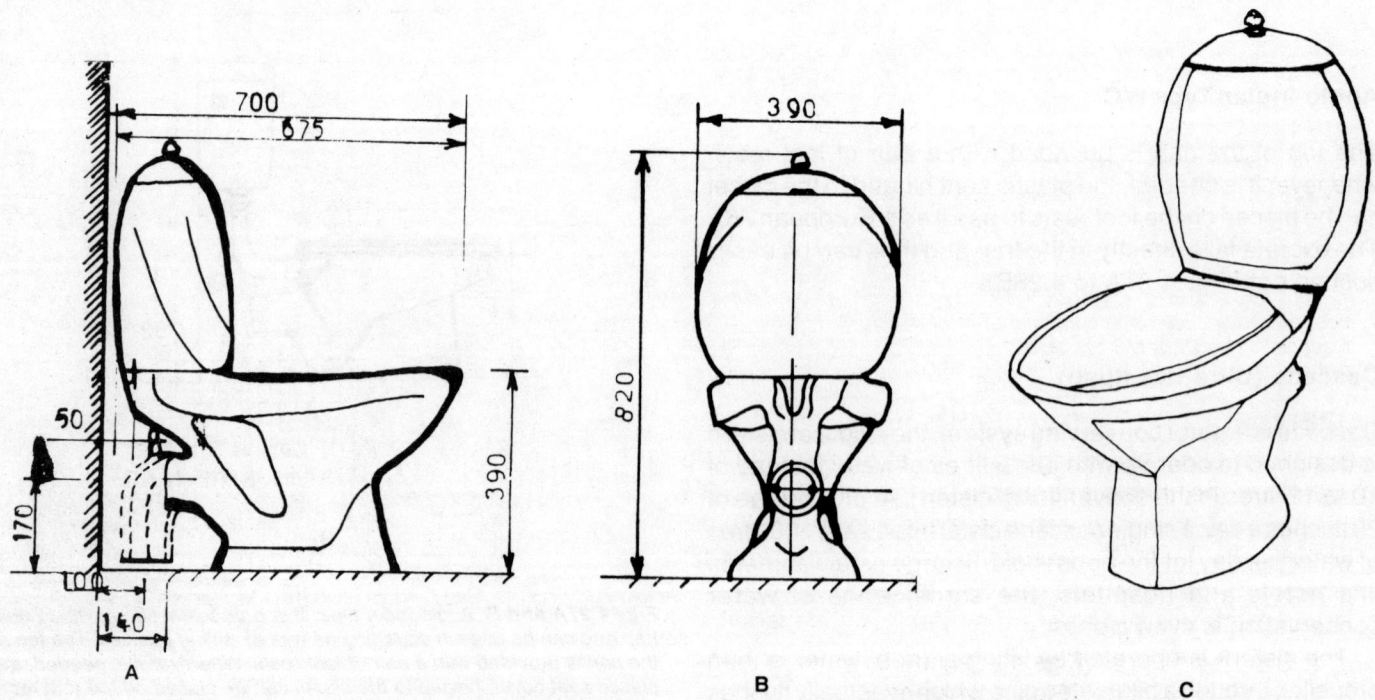

A

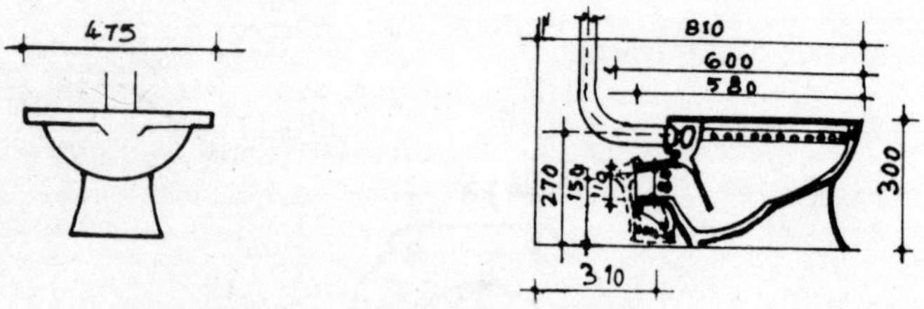

B

475

810
600
580

270 150 300

310

Figs 4.28A and B *Combination closet: Indian wash down water closet with a lower depth. It is also available in both P and S trap and can be used both with low level and high level cistern.*

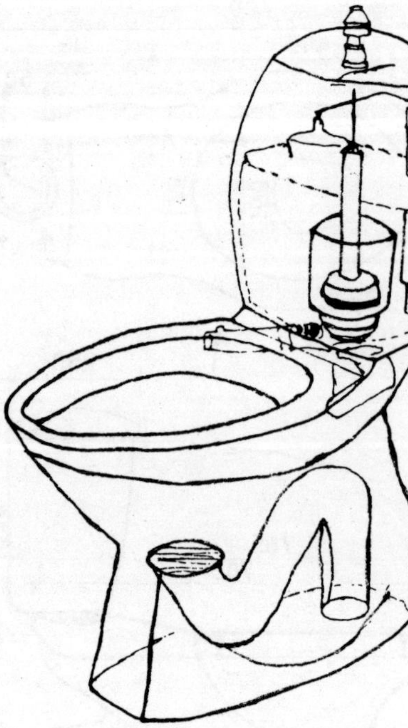

Figs 4.29 *In-built features of cascade type water closet.*

700
675

50

170

140

A

390

390

820

B

C

Figs 4.30A, B and C *Cascade water closet. It is available in both P and trap.*

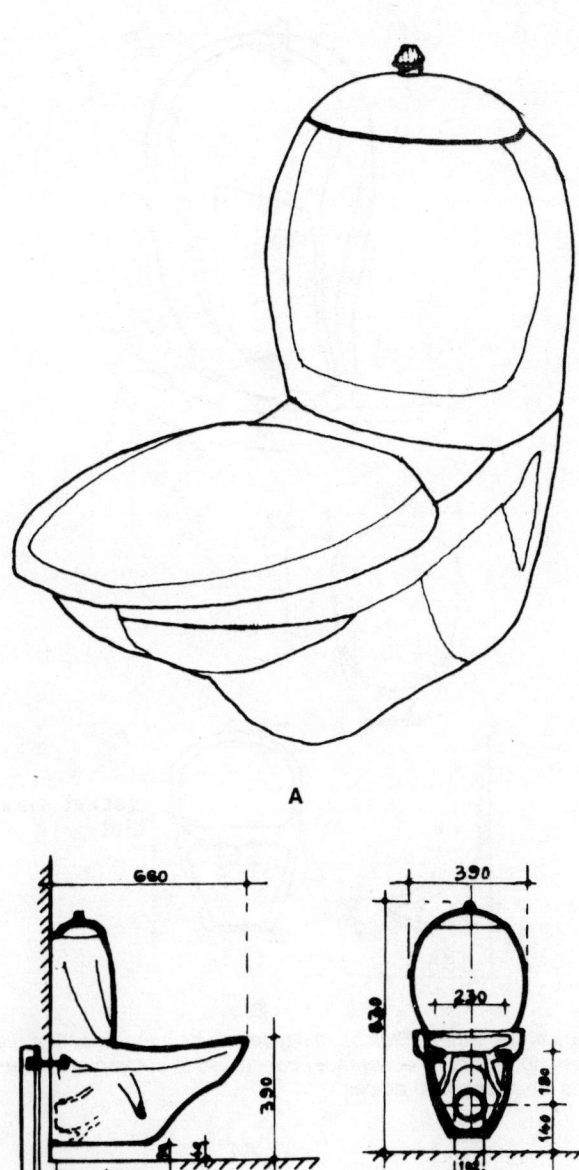

A

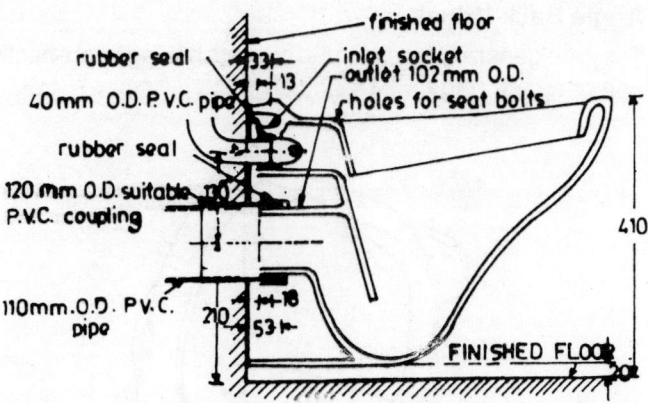

Fig. 4.32 Fixing arrangement of a cascade wall hung water closet. • If the wall is going to the plastered for a thickness of 6/7 mm. The projection of the bolts form the finished wall should be 66 mm only • The bracket should be buried into the floor. So that the height from the centre line of the bolt to the surface of the finished floor should be 320 mm.

Bowl-type Urinal

Bowl type urinal is with flat back, bowl type urinals are generally installed in single, without the requirement of partitions (Fig. 4.33A, B).

B   **C**

Figs 4.31A, B and C Cascade wall hung closet: This is the wall-mounted water closet with concealed P trap. This type of water closet is supported on CI or MS concealed brackets with bolts sunk into the wall and floor.

URINALS

Urinals fall under the category of soil appliances and the outlet from the urinals is connected to the soil pipe. Urinals are generally provided with automatic flushing systems. An antisiphonage pipe is necessary for urinals located on different floors and connected to a common soil pipe.

Various types of urinals are available in the market.

1. Bowl-type urinal (Figs 4.33A and B)
2. Angle back urinal (Figs 4.34A and B)
3. Slab or stall type urinal (Figs 4.35A to C)
4. Squatting urinal (Figs 4.36A and B)

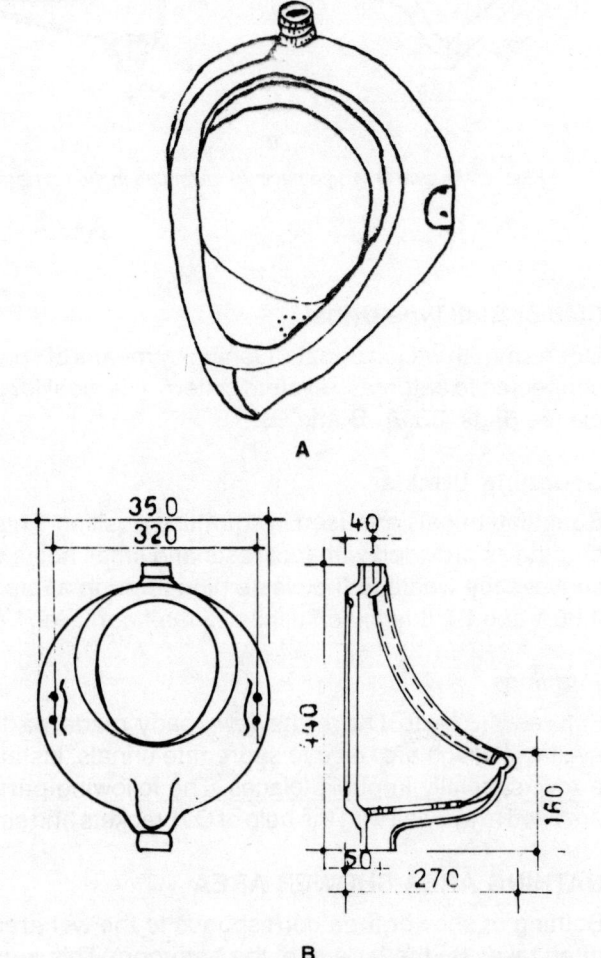

A

B

Figs 4.33A and B Flat back urinal with integral flushing rim.

Angle Back Urinal

They are generally for use in the corners having best effective use of space (Figs 4.34A and B).

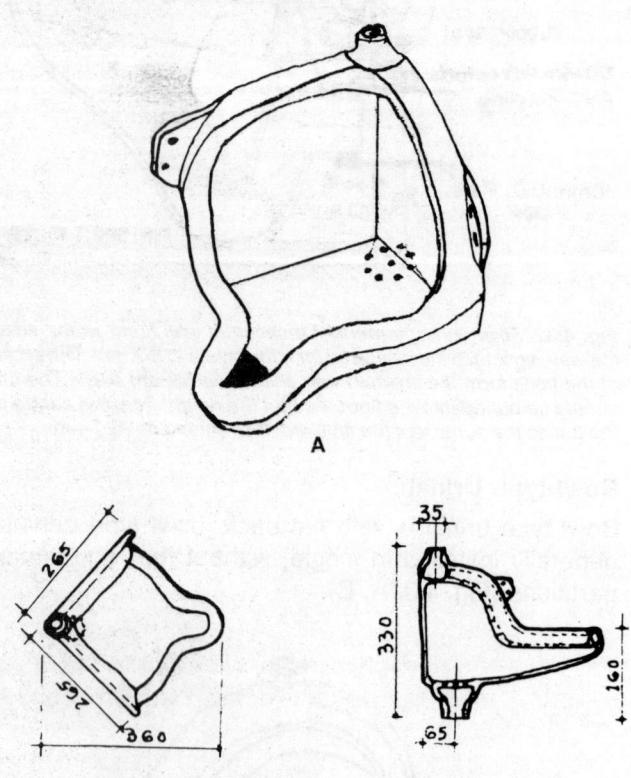

Figs 4.34A and B *Angle back urinal for use in the corners.*

Slab or Stall Type Urinal

Open smooth walled urinals flushing by means of spreader connected to automatic system cistern. It is ideal for public places (Figs 4.35A, B and C).

Squatting Urinals

Squatting urinals are used in squatting position. This type of urinal is provided with foot rests and small holes which conveys the waste to the waste pipe through a trap (Figs 4.36A and B). It is fixed flushed to the floor level.

Partitions

Figures 4.37A to D are the few ready-made partitions available which are used to segregate urinals, installed in a row, specially in public places. The following partitions are fixed to the wall with the help of CJ brackets and screws.

BATHING AREA/SHOWER AREA

Bathing or shower area corresponds to the wet area and often takes up the far end of the bathroom. This area can take any of the following forms (Figs 4.38A to C).

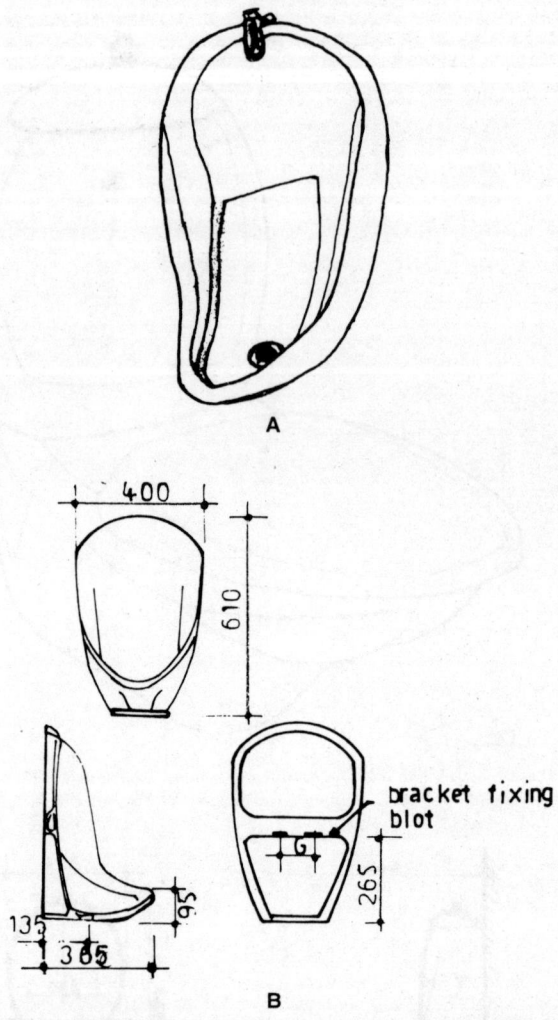

Figs 4.35A and B *Slab or stall type urinals open smooth walled urinal flushing by means of spreader connected to automatic system cistern. It is ideal for public places.*

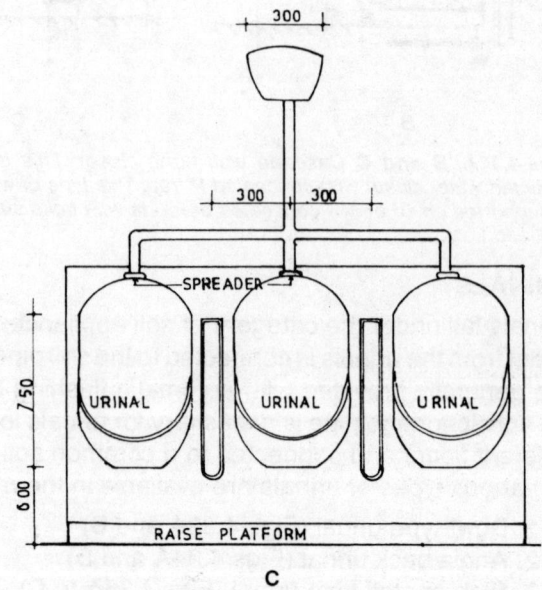

Figs 4.35C *Stall type urinals. The sketch shows the unit of three urinal in a row for use in public places.*

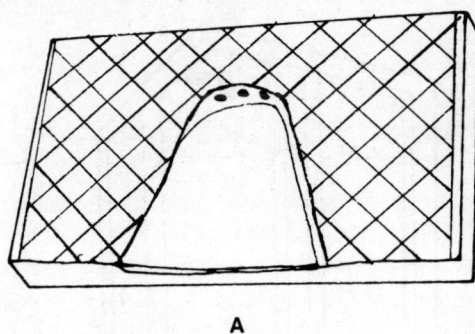

A

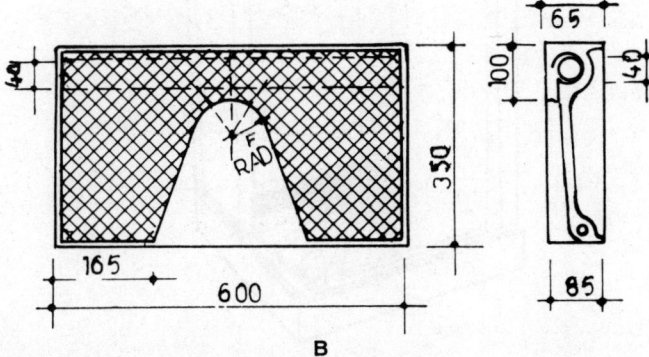

B

Figs 4.36A and B Shows squatting urinal used in squatting position. This type of urinal is provided the foot rests and small holes which conveys the waste to the waste pipes through a trap. It is fixed flushed to the floor level.

(i) Small tile lined corner of the bathroom or a separate room, complete with drainage. Shower tray of fibre glass or acrylic with slope and a drainage hole is available in the market, which covers the flooring of the area.

(ii) A simple shower attachment at one end of the bathtub, with wall mounted sockets or a slide bar enabling the handset to be fixed in a suitable position for easy use. This can have a curtain, glass or plastic screen to make it an enclosure to avoid the water splashes outside the bathtub.

(iii) A shower cubicle either specially built for its site or comprising one of the ready-made cubicles which are widely available. This will either have a fixed shower head or an adjustable handset.

The following points should be remembered while designing the shower.

(i) The drainage should be sufficient with floor sloping towards the drainage hole so that one does not have to wipe the floor after each bath with a mop.

(ii) The water outlets should be adjusted to suit the needs of various sized users. A fixed shower head can simply be hinged so that its direction is changed as required, or the hose and handset type can move up and down a slide bar, or have a telephone shower with a flexible hose.

(iii) A seat is a splendid addition, particularly for elderly people, and the shower head should be arranged so that it can be directed over the seat.

(iv) If the shower is in a fixed position, the plumbing can be concealed in the wall with the controls at a convenient height.

(v) Provide soap holders, grab handles and towel rails within easy reach and at the right height for everyone using the shower.

(vi) Thermostatic shower controls if used, ensure that water is at the correct temperature without the need to balance the hot and cold water supply to the shower through the taps. They compensate

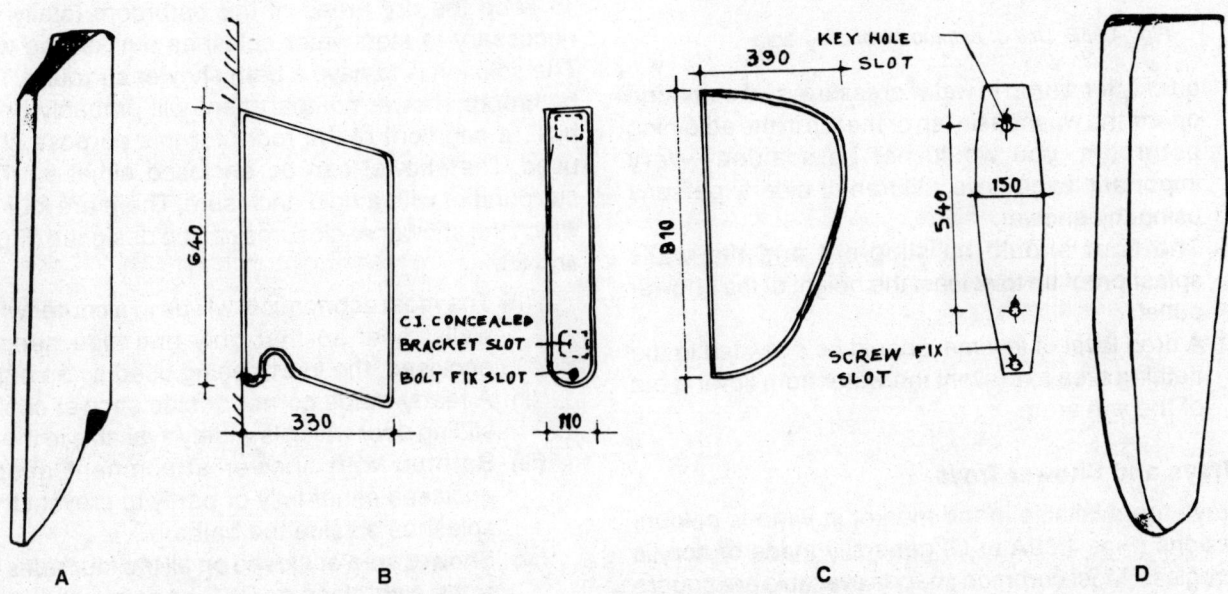

A   B   C   D

Figs 4.37A, B, C, and D Segregations used for urinals at public places.

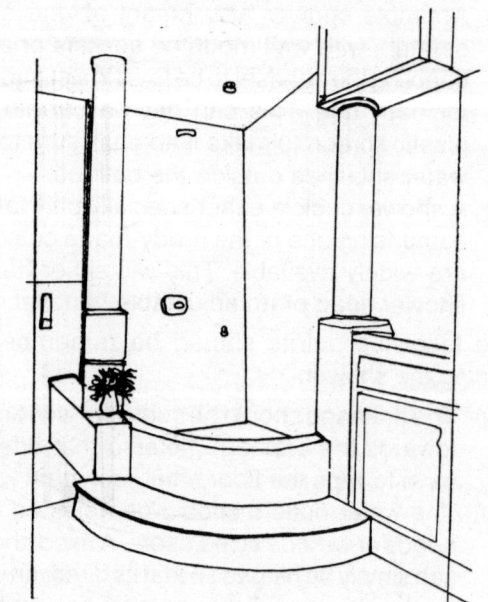

Fig. 4.38A *Sketch of enclosed sunk in shower area.*

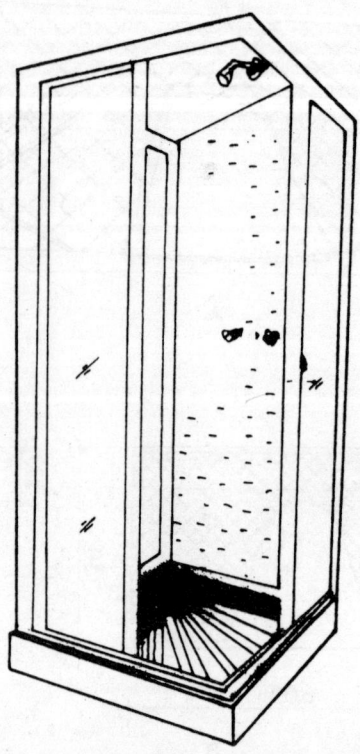

Fig. 4.38C *Use of shower tray as enclosed bathing/shower area.*

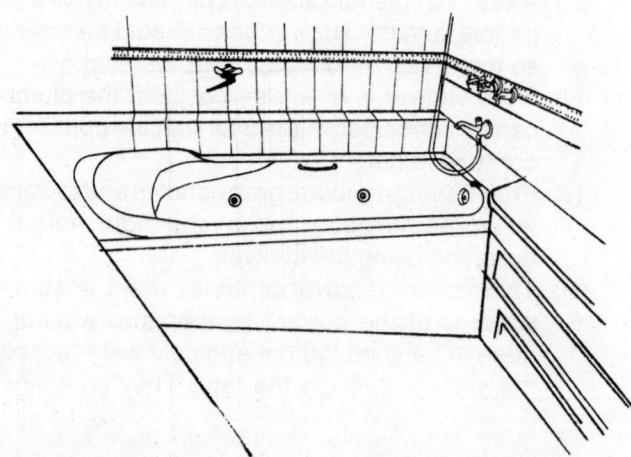

Fig. 4.38B *Use of bathtub as bathing area.*

quickly for a drop in water pressure, so if someone opens the washbasin tap or the tap in the adjoining bathroom, you would not be scalded—very important if you have children or elderly persons using the shower.

(vii) The floor should be slipproof and the walls splashproof up to at least the height of the shower outlet.

(viii) A drop level of few mm should be provided in the bathing area to prevent the water from flowing out of the wet area.

Bath Trays and Shower Trays

Bath trays are available in the market in various colours and designs (Figs 4.39A to C) generally made of acrylic and fibreglass. Most common shapes available are square and rectangular and in various sizes to fit in various space requirements. The depth of these trays is generally 125 mm to keep the water splashes inside the bathing trays only. These trays have sloping floor and a hole for proper drainage and have slipproof floors. The range starts from 760 mm × 760 mm for square trays and 900 mm × 600 mm for rectangular shapes. Also available are the trays with a raised seating space and place for soap holders, etc. The depth of these trays is generally 125 mm.

Shower Enclosures

To keep the dry areas of the bathroom totally dry, it is necessary to stop water splashes outside the wet area. The solution is to have a bath shower surround. The ideal bathroom shower compartment will probably be custom built, a segment of the room for this purpose should be used. The shower can be enclosed either by the fabric surround or with a rigid enclosure. There are four ways by which the shower enclosures can be designed (Figs 4.40A and B).

(i) The most economical will be in a corner where two walls meet so that only one side needs to be enclosed, the fourth being used as an entrance.

(ii) A ready-made corner or side shower cubicle with sliding door which is widely available in the market.

(iii) Bathtub with shower attachment need to be enclosed either fully or partly to prevent the water splashes outside the bathtub.

(iv) Shower area enclosed on all the four sides or three sides with glass or PVC panels, either sliding or hinged.

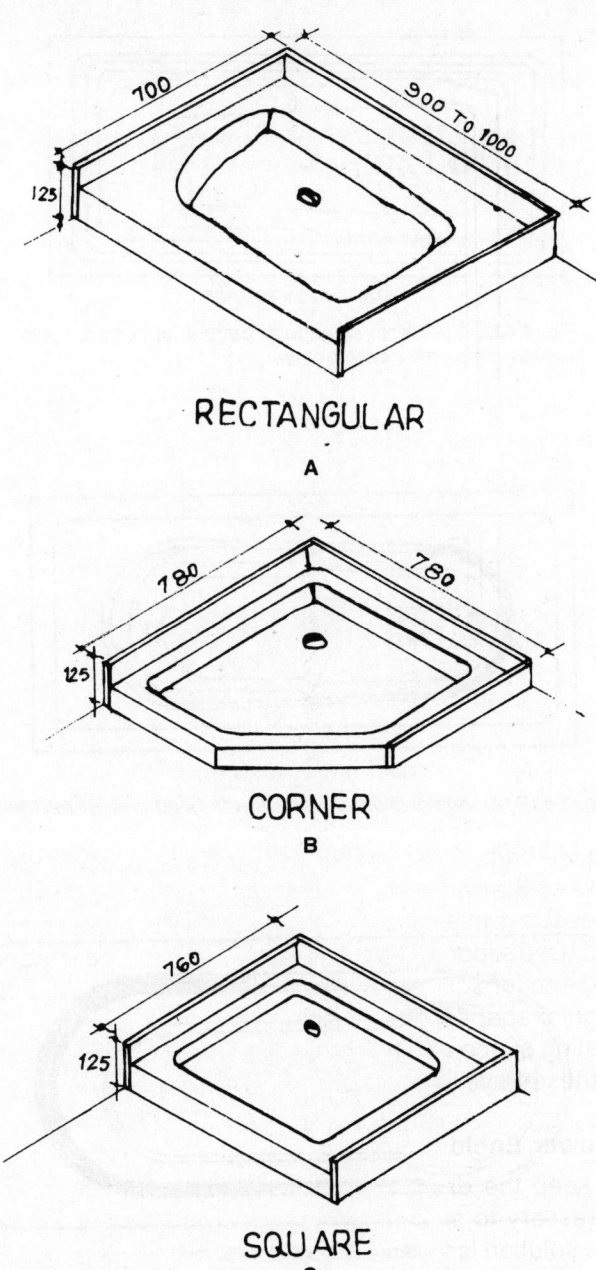

Figs 4.39A, B and C Different type of shower trays.

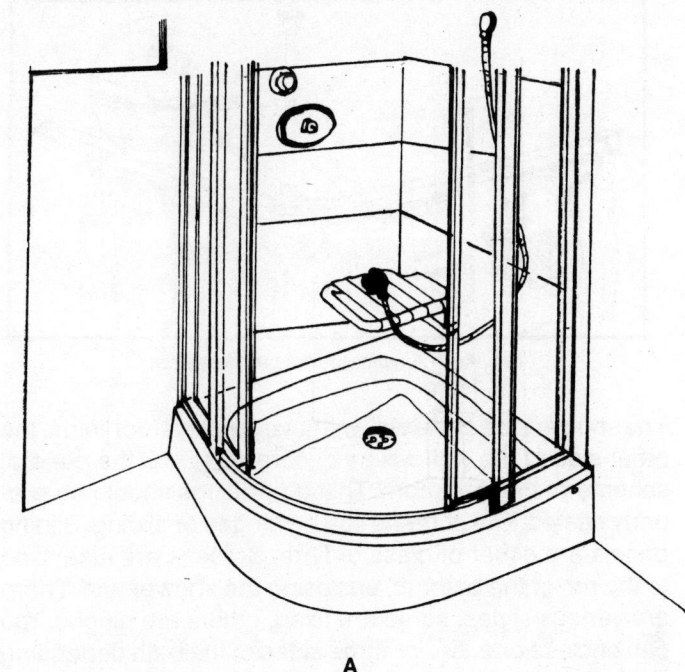

A

B

Fabric Curtains and Screens

These are inexpensive and easy to match in, but often less good at containing the spray.

Shower curtains are hung from a continuous rail fixed to wall and/or ceiling (Fig. 4.41). Curtains allow you to enclose as much of the bath as you want, depending on the track layout. Where there is a wall at each end, a straight rod can be fixed. For L, U and square-shaped tracks, wall brackets and rail brackets can hold the rail.

Partition

The partition wall can be of brick, PVC, concrete blocks, glass brick, timber or any other material, or it can be a simple timber or PVC frame with plastic or glass panels.

Figs 4.40A and B Typical shower enclosures with sliding partitions.

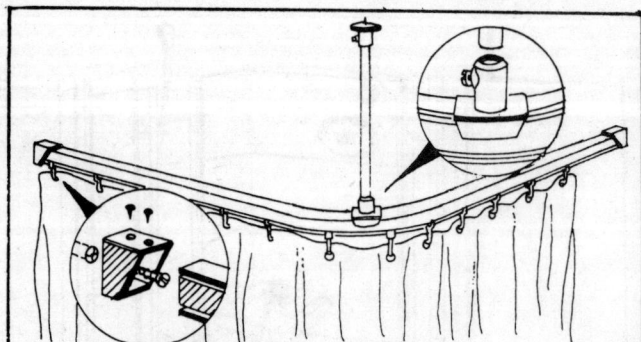

Fig. 4.41 *A typical curtain track system.*

The shower side of the wall will have a waterproof lining, the other side of the wall will be decorated to suit the general scheme of the bathroom. These partitions should be properly sealed. The screens can be hinged or sliding. Sliding panels are either of glass or fibre. Screens are also fixed to the rim of the bathtub, enclosing the shower end. There are various styles, some are fixed, others are hinged. You can enclose one, two or three sides of the bath depending on the layout of the bathroom. The bottom of the screen is properly sealed with rubber sealing strips (Fig. 4.42).

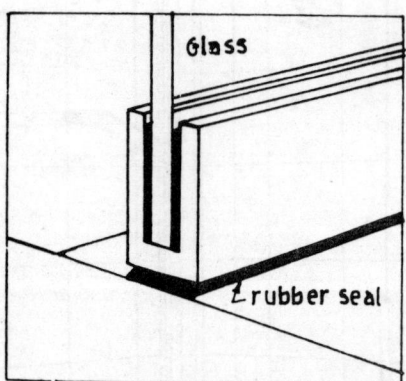

Fig. 4.42 *Details showing aluminium sliding frame with glass/fibre glass panel.*

BATHTUBS

Use of bathtubs is becoming quite popular these days. Bathtubs may be precast or cast-*in-situ*. They are made of enamelled steel, gel-coated fibre glass, enamelled porcelain, moulded plastic or reinforced concrete finished with terracotta or marble finishes (Colour Plate 5). It is provided with outlet or overflow piples which are usually of 40 mm diameter. Provision is kept for both hot and cold water connections.

The most common type of bathtubs that are used are acrylicopaque coloured 5 mm thick acrylic sheets reinforced with fibre reinforced plastic (FRP)—moulded in the shape of a bathtub. Bathtubs are available in various shapes like oval, round, rectangular or corner tub. Following are few of the common types available in the market under different trade names (Figs 4.43A to N).

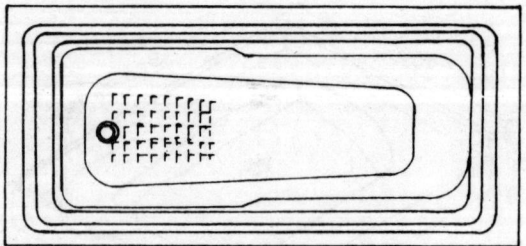

Size - 1525 × 725 × 370

Fig. 4.43A *Standard rectangular bathtub of compact size suitable for smaller bathrooms.*

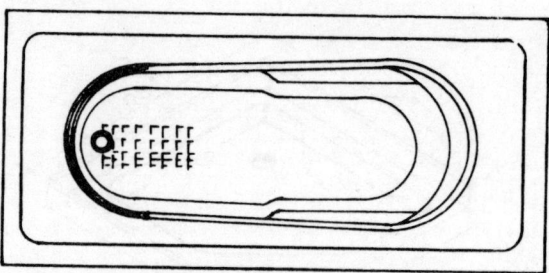

Size - 1670 × 790 × 370 mm

Fig. 4.43B *Standard shaped bathtub with head rest for relaxation.*

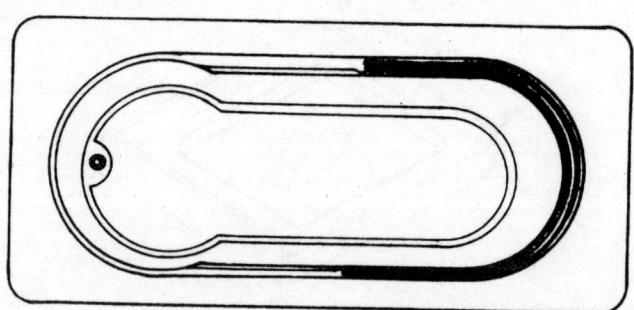

Size - 1790 × 890 × 450 mm

Fig. 4.43C *Standard rectangular bathtub with additional interior depth.*

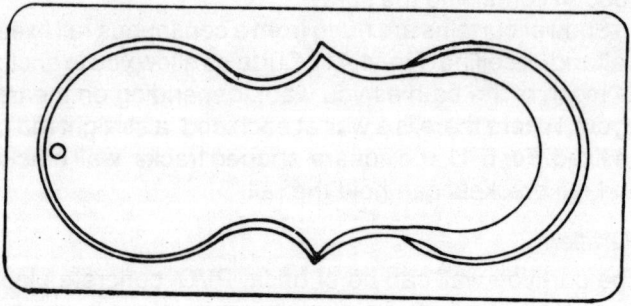

Size - 1790 × 890 × 450 mm

Fig. 4.43D *Standard rectangular bathtub with rounded interior.*

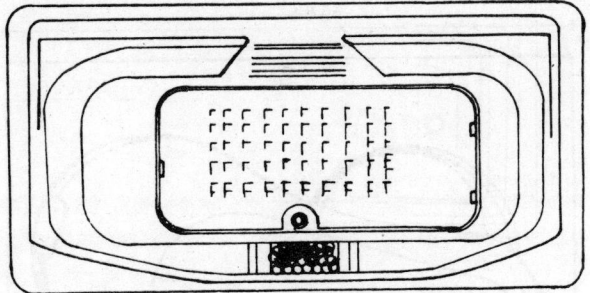

Size - 1700 × 850 × 400 mm

Fig. 4.43E *Standard rectangular bathtub having drainage hole on the longer side with built in arm and back rest.*

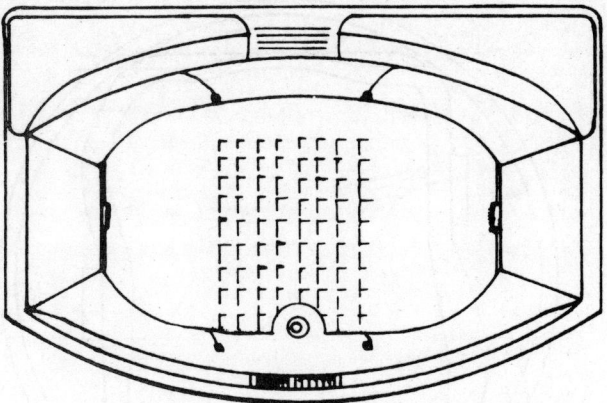

Size - 1780 × 1200 × 425 mm

Fig. 4.43H *Fancy bathtub rounded curved shape.*

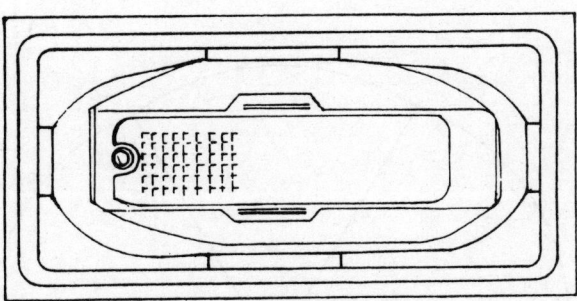

Size - 1700 × 850 × 400 mm

Fig. 4.43F *Bowl shaped bathtub having sloping back rests and grab bars.*

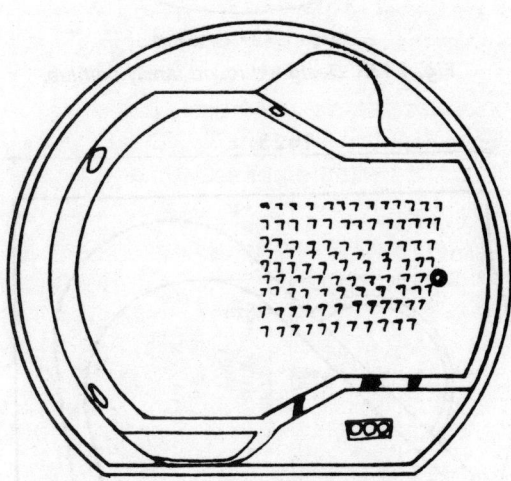

Size - 1580 × 1380 × 400 mm

Fig. 4.43I *Moon-shaped bathtub designed to accommodate a couple/two bathers. Countered back rest and built in arms rest for relaxation.*

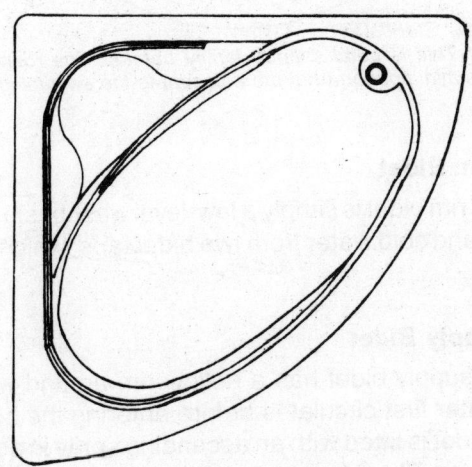

Size - 1400 × 1400 × 450 mm

Fig. 4.43G *Corner bathtub suitable for utility cum fancy bathrooms.*

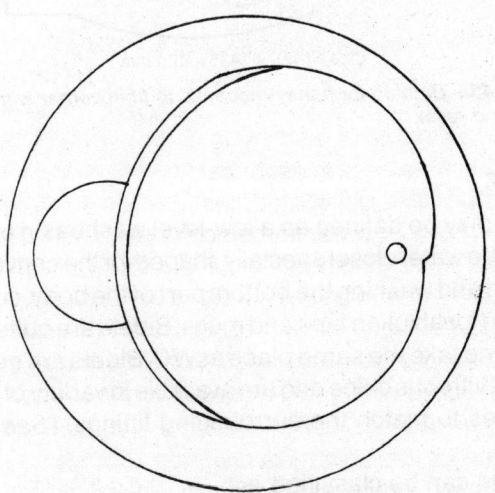

Size - 1500 ⏀

Fig. 4.43J *The gently rounded bathtub with a lounger seat.*

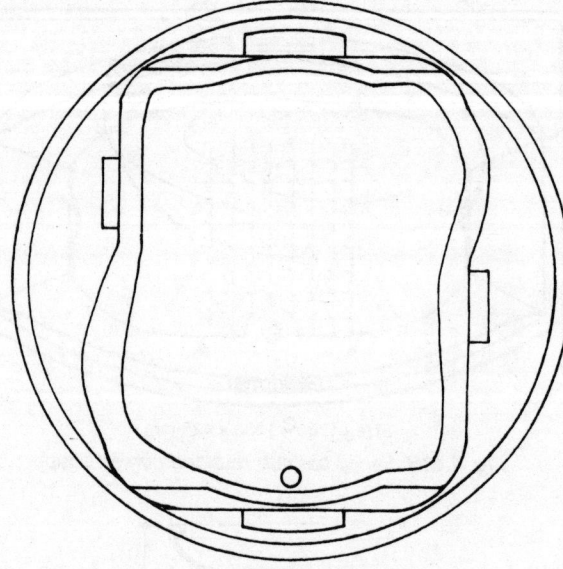

Size - 1740 ⌀ × 450 mm

Fig. 4.43K *Designed round family bathtub.*

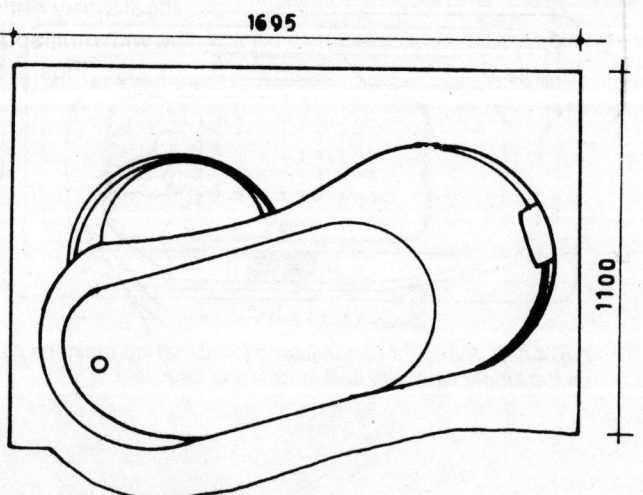

Size - 1695 × 1100 × 400 mm

Fig. 4.43M *Mother and child bathtub with a cushioned head rest for rest and relaxation.*

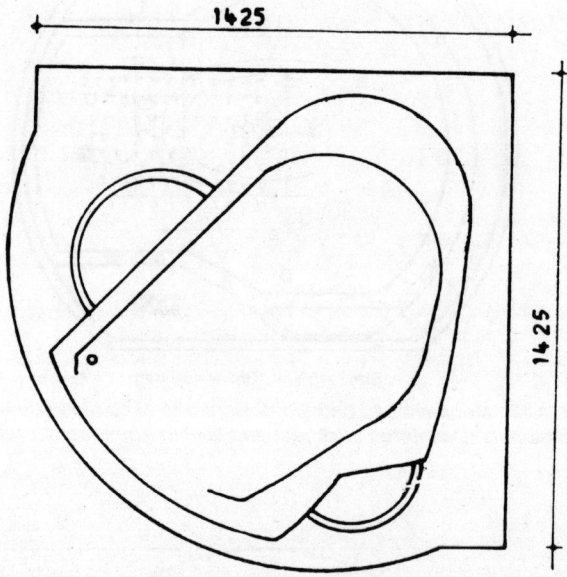

Size - 1425 × 1425 × 425 mm

Fig. 4.43L *Bathtub for honey mooner's to fit in corner with lovers seat and rests.*

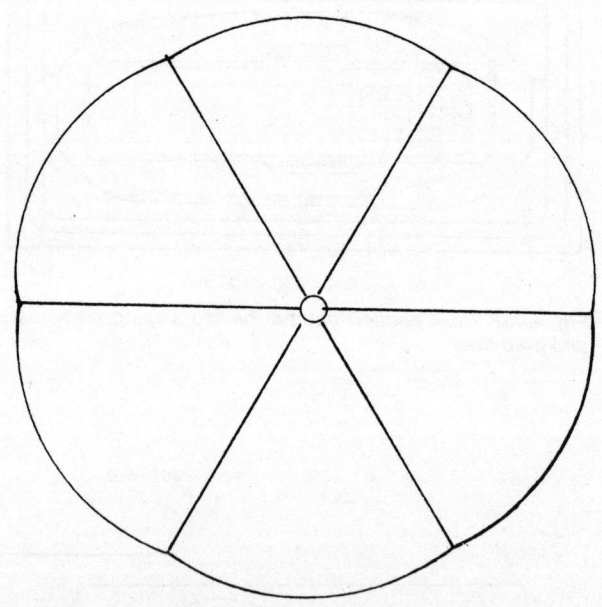

Fig. 4.43N *This is petal shaped family bathtub. The rounded petals function as mini pool or a hot tub adjacent to the swimming pool.*

BIDETS

A bidet may be defined as a low-level washbasin provided next to the water closet specially shaped for the convenience of sitting and washing the bottom part of the body, curtailing the need for ablution taps and mugs. Bidets are quite similar to WC and take the same place as WC. Bidets are generally made of vitreous china and are available in variety of shapes and sizes to match the surrounding fittings. (*See* Colour Plate 4).

Bidets can be classified as:

(i) Over rim bidet
(ii) Rim supply bidet.

Over Rim Bidet

The over rim bidet is simply a low-level washbasin supplied with hot and cold water from two bidet taps, which gets the water.

Rim Supply Bidet

The rim supply bidet has a hollow rim around which the warm water first circulates before entering the bowl. This type of bidet is fitted with an ascending spray in addition to the bidet tap. The advantage of this type of bidet is that the circulation of warm water in the rim warms the rim, making it more comfortable to sit. At the flick of control knob the

warm water is diverted from rim to the vertical spray outlet
at the bottom of the bowl. Bidets are available with multispray
nozzle and pop-up waste outlet arrangement also (Figs 4.44
to 4.47).

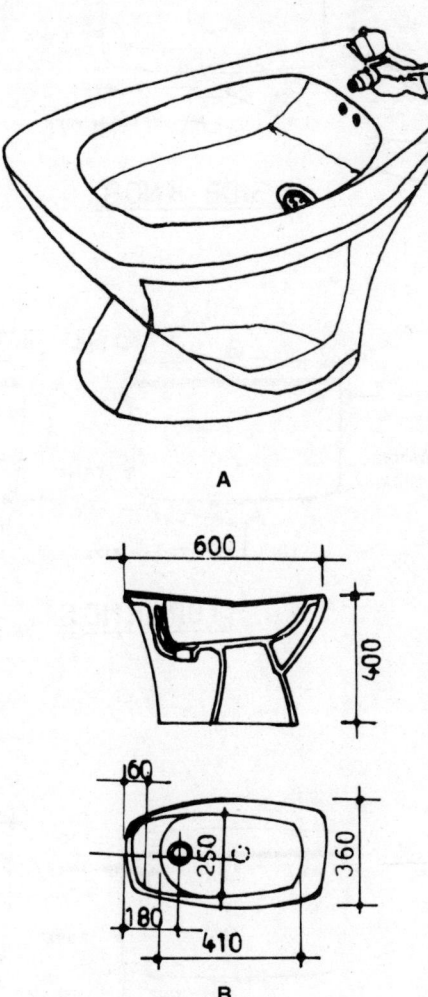

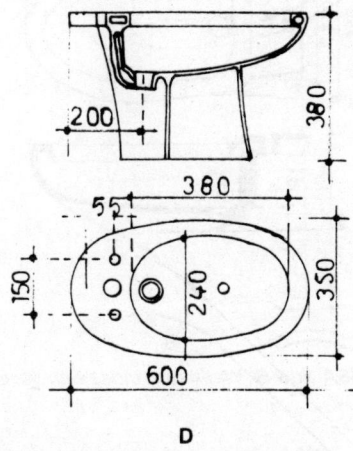

Figs 4.44C and D *Four hole bidet with vertical spray
having self-warming integrated flushing rim.*

Figs 4.44A and B *Floor-mounted bidet with
single tap hole without vertical spray.*

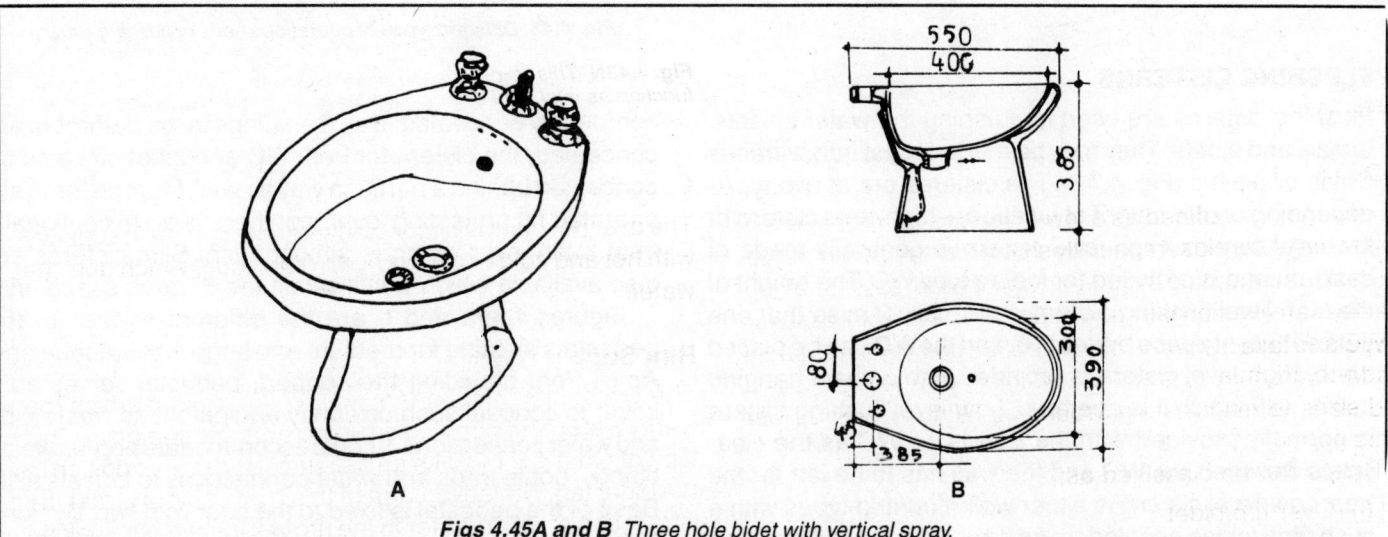

Figs 4.45A and B *Three hole bidet with vertical spray.*

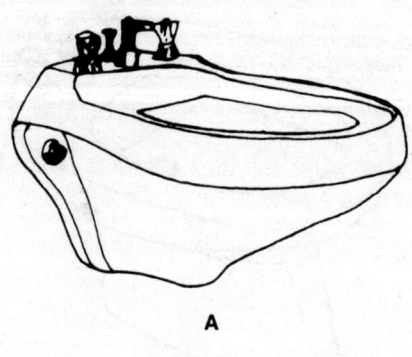

A

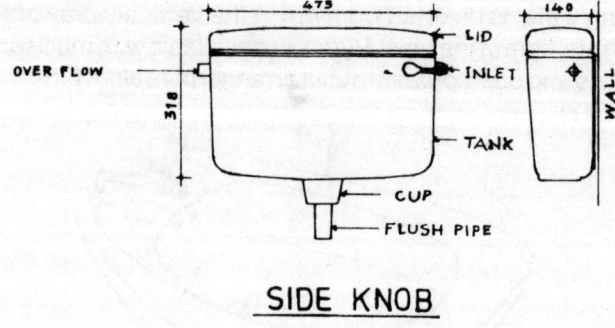

SIDE KNOB

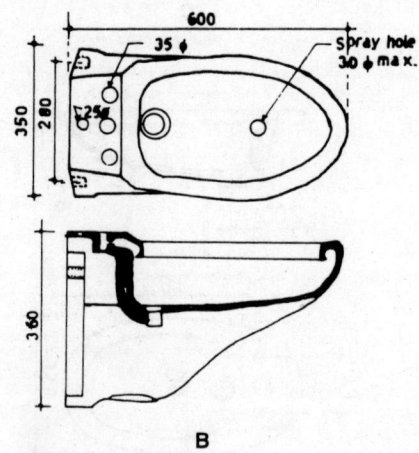

B

Figs 4.46A and B *Wall-hung bidet with vertical spray.*

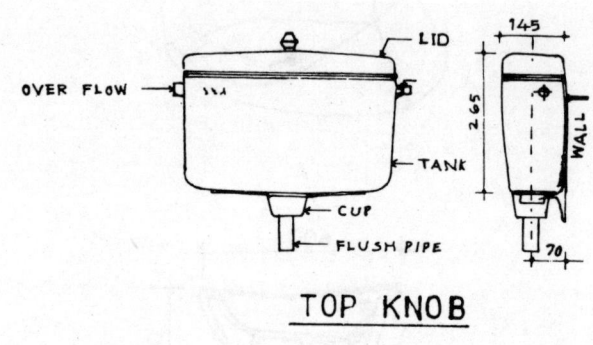

TOP KNOB

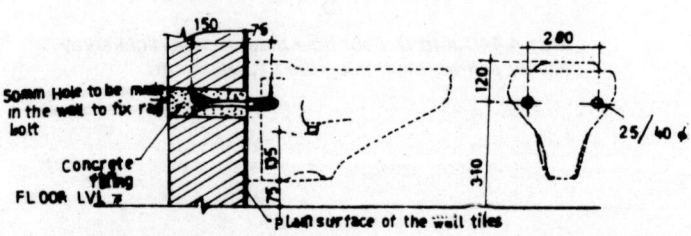

Fig. 4.47 *Fixing arrangement of wall-hung bidet.*

CHAIN HANDLE SYSTEM

Fig. 4.48 *Different types of operation knobs of flushing cistern.*

FLUSHING CISTERNS

Flushing cisterns are used for flushing the water closets, urinals and bidets. They may be made of cast iron, vitreous China or plastic (Fig. 4.48). The cisterns are of two types depending on the height of location—high-level cistern or low-level cistern. High-level cistern is generally made of cast iron and is provided for Indian type WC. The height of the high-level flushing cistern is normally 2 m so that one gets sufficient space beneath it, and the WC can be placed there. High-level cistern is provided with a chain hanging down, with which it is operated. Low-level flushing cistern is normally provided with the European WC as the clear space between the WC and the wall has to be left for the trap. Low-level cistern is either wall-mounted types with a flush pipe, close coupled to the pan, or best of all it can be

completely concealed. If all the fittings in the bathroom are concealed, the cistern for both WC and bidet can also be concealed behind a partition wall as well. Most cisterns are operated by projecting lever, but there is an exceptionally neat push lever which is almost flush. Slim cisterns are also available which will fit within the 6" deep space only.

Figures 4.49a and b are the different shapes of the pedestals suitable for medium and large size washbasins. Apart from providing the support, pedestal serves as a cover to conceal such unsightly equipment as waste pipe and water connections. It could accommodate pop-up waste fittings, bottle traps and water connections to faucets also. Base of the pedestal is fixed to the floor with two 1½" long wood screws.

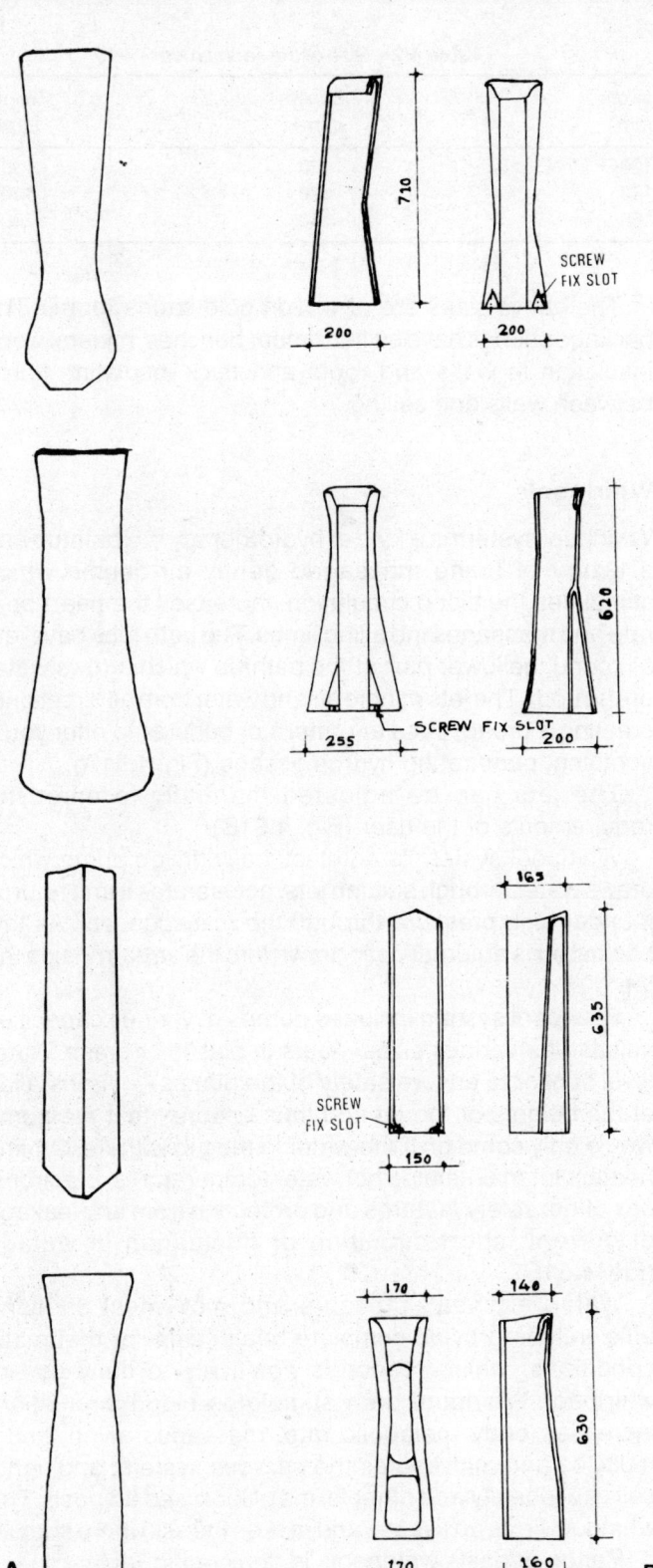

Fig. 4.49A, B *Different types of washbasin pedestals.*

LUXURIES FOR THE BATHROOMS

Depending on the environment you have decided to create, bathroom extras can constitute anything from a mink-covered pouffe to a sauna, with a whole range of health and relaxation equipment in between. One which could be regarded as practical rather than an extravagance is a telephone extension, the comforts of a steaming tub can add immeasurably to the pleasures of a friendly conversation, and take the sting out of a business call. Phone points do not react well to condensation, however, so ensure that the installation is protected by a waterproof shield.

Sauna/Steam Bath

The technique of sauna/steam bathing can be traced back to the ancient past. The Romans were well acquainted with the health benefits of steam, and built elaborate bathing complexes around the natural hot springs of their empire. The custom of Turkish bath recognises the benefits of damp heat and excessive perspiration to rid the body of toxins.

Heat—damp or dry—is one of the best body cleansing techniques known and it also stimulates blood circulation. As with the hot tub or furo, the process of soaping, scrubbing and rinsing traditionally precedes and follows the sauna or steam bath. But they are two different techniques that require different types of equipment.

The Steam/Turkish Bath (A low temperature high humidity bath in a glass fibre room)

The steam bath is a form of water treatment with hot steam and is confirmed to have relaxing and detoxifying action dilating the blood vessels with a beneficial effect on the head. It plays an important role in relieving problems such as diseases of the respiratory organs. Steam bath reduces the level of fat in the body leaving a tingling feeling of health and vigour.

The steam penetrates deep into the skin, cleanses it, massages the skin, increasing blood circulation and easing tense muscles. The steam room is generally in concrete with tile or marble finished walls supported with steam generators, separate control panel and other accessories.

Steam generator has a stainless steel water reservoir with heating elements which produces steam continuously. It is provided with drainage pipe and sometimes steams head with a container for liquid essences. The unit can be located 15 m from steam room. Now-a-days automatic generators are available, which empty the reservoir one hour after completion of bathing. Control panel regulates bathing time and ensures an absolutely constant bathing temperature and can be located up to 4.5 m from the steam room.

Steam baths are also available as a packaged unit with built-in steam generator and control panel. These are tempered glass cubicles provided with the drainage cubicle size usually 120 cm × 120 cm × 200 cm.

The Sauna Bath (A high temperature low humidity bath in a wooden panelled room)

A sauna is a healthy exercise for capillaries and enables a thorough detoxification and the skin attains a healthy glow. The blood circulation regains strength, respiration improves, the organism defences are stimulated, rheumatic pains lessen or disappear. It provides an effective work-out for the cardiovascular and thermoregulation systems, combats excess weight, athletic performance.

In sauna, the skin is cleansed by intense perspiring, the pores are opened, and the muscles are relaxed. Relaxation is as important to the sauna as the alternating exposure to dry heat, and cold showers. A proper sauna needs only to have an area sufficient for the bathers to recline in. The construction of a sauna is basic and straightforward. A small room lined with wood and tight fitting door and double glazed windows and sauna heater are the necessities (Fig. 4.50). The choice of wood is specially important not simply for its appearance and fragrance, but to enable the bather to recline or sit comfortably when surface temperatures are high. A sauna should be further insulated with fibreglass. When available space does not determine the sauna size, a good yardstick is 2 ft of bench space per bather, while 5 ft by 7 ft, is the minimum space for a person lying down (can be settled by the number of people who will use it).

Fig. 4.50 *Details of sauna bath.*

The sauna can be healthy and worthwhile investment in bathroom and family enjoyment, and need not take much space or to be expensive to install—they are now available as a package purchase—a sauna bath cubicle which can be placed anywhere in the bathroom or bedroom as it does not need any drainage.

Of the kits that are available, some are no larger than the closet size. Rarely is a sauna larger than 12 ft wide for efficient heat circulation. Normally, sauna rooms come in the following sizes (Table 4.2).

Table 4.2: *Size of the sauna room*

Width (cm)	Length (cm)	Height (cm)
120	120	190
180	120	190
150	140	200

The above sizes are of woven gold sauna rooms. The package saunas have built-in timber benches, mineral wood insulation in walls and roofs and thick insulating fibres between walls and ceiling.

Whirlpools

Whirlpool system is a kind of hydrotherapy in a bathtub and a luxury of being massaged gently air-bubble which stimulates the blood circulation, increases the heart beat rate and massages the ailing limbs. The bath tubs have jets all round the lower part of the bathtub which throws water on the body. The jets mingle air and water from all directions, creating a broad, circular pattern of bubbles to offer you a soothing, penetrating hydromassage (Fig. 4.51A).

The jets can be adjusted manually to meet the requirements of the user (Fig. 4.51B).

Whirlpool system is an electrically driven pump which draws water through suction jets, accelerates it and returns it under high pressure through the massage jets. As this occurs, air is automatically drawn into the stream inside the jet.

Whirlpool system includes pumps of various capacities suitable for various sizes—Jets in plastic or brass. Water level sensor to ensure safety of the pumps—it is installed at the heights of the jet and thus ensures that the pump would only come on if the water in the tub is there. On-line heaters for maintaining hot water temperatures. Electronic box offers safety features and protection from any leakage of current, short circuiting or fluctuation in voltage (Fig. 4.51C).

Water, air, heat, pressure and movement combine efficiently to provide complete health care for rheumatic conditions, which responds positively to daily use of whirlpool. Whirlpool bath stimulates blood circulation, increases body metabolic rate, massages ailing limbs/muscles and rejuvenates the nervous system, and fights cellulite, obesity and other forms of localised adiposis. The whirlpool firms up tissues and makes the skin more supple.

Range of oasis whirlpools is depicted in Table 4.3.

Table 4.3: *Range of oasis whirlpools*

Max. jets	Flow rate	Suction	Delivery	HP	Amp
6	430 L/min	50 mm	40 mm	1.0	5.1
8	560 L/min	50 mm	50 mm	1.25	5.5
10	650 L/min	50 mm	50 mm	1.5	6.0

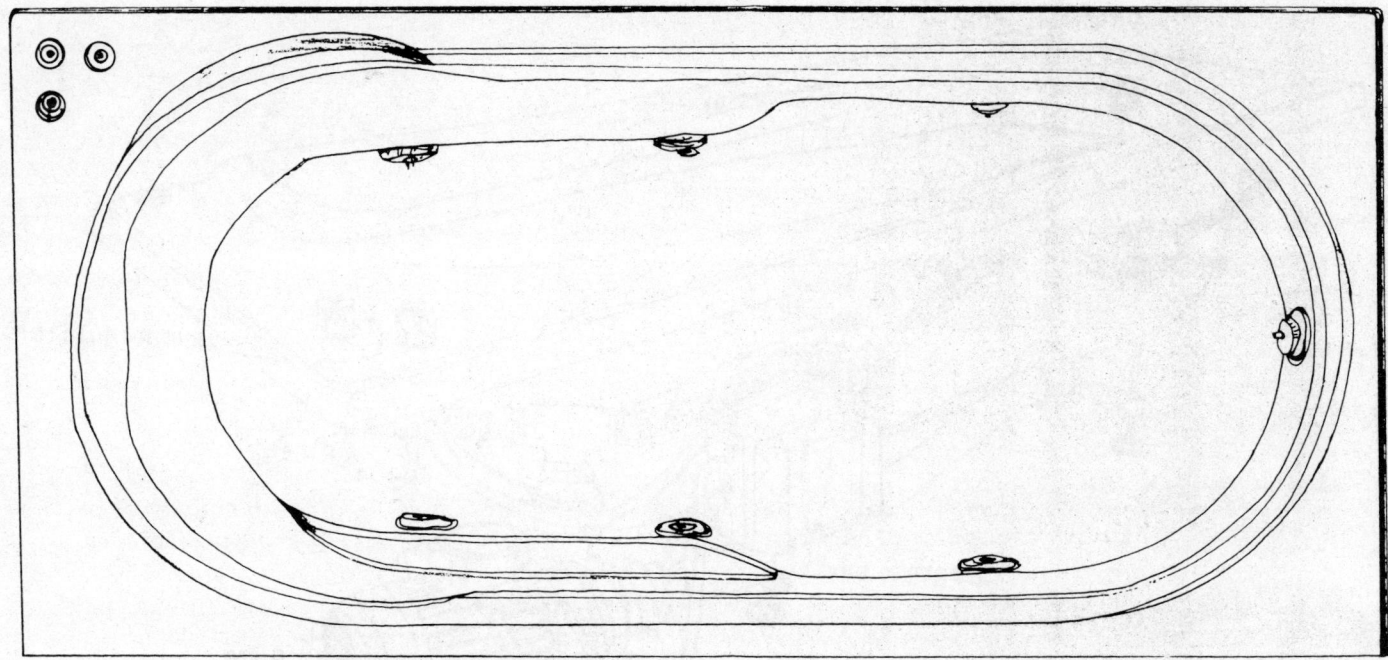

Fig. 4.51A *Plan of whirlpool/jacuzee tub.*

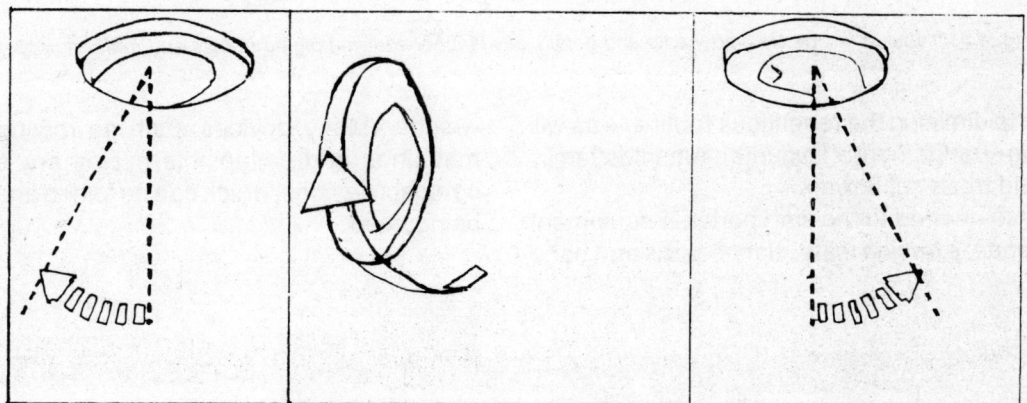

Fig. 4.51B *Details of adjustable jets in whirlpool.*

Excercise Bathrooms

In ancient Greece, bathing was a part of the gymnastic and exercise routine. Today the availability of more compact exercise equipment and home saunas, steam rooms and whirlpools means that an exercise routine can be more easily integrated into the home, most likely in the bathroom. For many, the privacy and convenience of home area is the only way to keep up an exercise routine. This may be no more than a warm up and stretching area, or a place for yoga, or it can be year-round exercise health spa.

Regular exercise can have several related benefits: it improves aerobic health, adds to muscular development, contributes to weight control, and lowers physical stress and tension. The routine and the types of equipment in a home gym should be chosen with personal exercise goals in mind.

Muscular strength is increased through exercises in which the muscles work against some resistance, e.g. lifting weights, push-ups and sit-ups. This type of exercise can need a little more than a mat. Relaxation or the reduction of muscular tension is probably the one form of exercise most easily accommodated by any bathroom. A hot bath can achieve this, but of course whirlpool tub, steam bath, sauna bath are far more effective and enjoyable.

An excercise area should have a soft, non-skid floor or be covered with an exercise mat with a slip cover of canvas or other natural fibre. Moderate, uniform lighting is best. Good ventilation is necessary, both to ensure plenty of air circulation and control odours.

The surroundings for exercise exert a powerful influence on the enjoyment, frequency and therefore, the results.

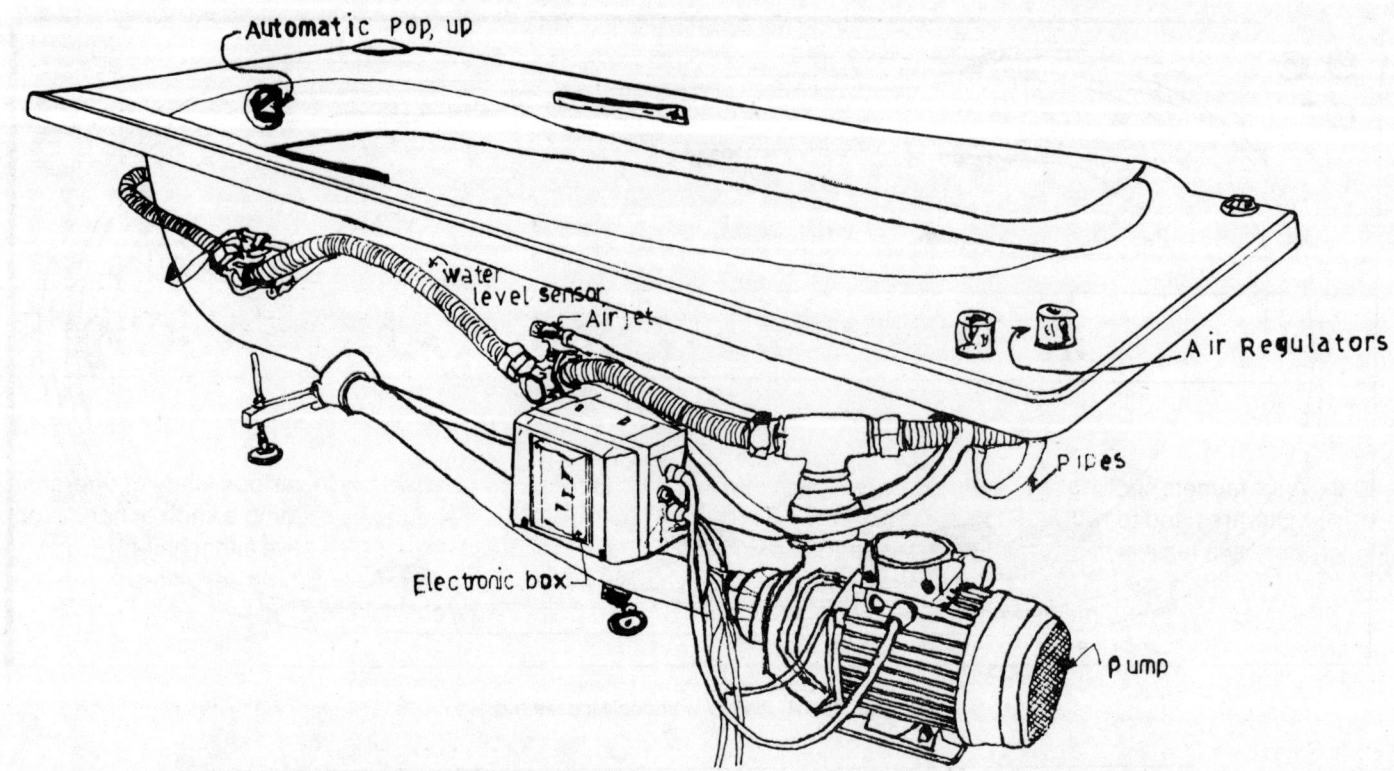

Fig. 4.51C *Inside technical components and details of whirlpool showing piping details and motor pumps, etc.*

Views can help to diminish the repetitious routine—as will music or evolving arrangement of favourite momentos, family photographs and fresh cut flowers.

Storage is also needed for holding portable equipment such as jump ropes, exercise mats, slant boards and hand weights. Many devices such as rowing machines, ski-machines and collapsible cycles are also available in compact designs, which can be folded and stored when not being used.

BATHROOM FITTINGS AND FAUCETS

Bathrooms faucets such as taps, stopcocks, showers, flush valves, etc. are used to regulate the supply of water. These faucets are available in the market in various materials, like chromium-plated metal, brass, powder-coated metal, aluminium or synthetic metal (symet). These faucets are available in various shapes, sizes and designs to suit various needs. The faucets and fittings available range from simple basic necessities to ultraluxury items. (*See* Colour Plate 6).

TAPS

Taps are the most important amongst the bathroom faucets. They can be either wall mounted or fixed on a horizontal surface such as bath or basin rim. The choice of tap is mostly a matter of appliance type or preference. Wall-mounted taps leave the rim clear and are easier to clean. The stem of the tap should be long enough to avoid awkward areas under the spout, which are difficult to clean and can interrupt the water flow. Bath/shower taps are normally wall mounted. Basin taps should be as unobtrusive as possible so that they do not interfere with hand and face washing in limited activity area. They should have long stems and projecting lever handles. All taps should operate smoothly throughout their opening and closing range, and tap handles should be comfortable to use.

The taps are available with various kinds of operations. They are either operated by rotating a knob or handle or by single lever. Taps with handles have either half-turn, quarter-turn or full turn operation. Single lever taps are much convenient to use as it is operated by simply changing the direction of a lever. There are special kinds of faucets, too: (i) the bathroom faucets are available with aerators to avoid the splash of water—the aerators filter the water and mixes air-bubbles with it, so that one does not get splash, and (ii) the taps are available with pop-up-drainage system for bathtubs and bidets. You simply have to pull the lever at the back of the tap to drain out the water from the bathtub and bidet after use.

The following are the basic types of faucets available in the market.

As mentioned earlier, there are basically two types of taps—one with the vertical inlet, and other with the horizontal inlet and are known as pillar cocks and bib cock respectively.

Pillar Cocks

Pillar cocks are the taps with vertical inlet, i.e. can be fixed on the horizontal surface of bath or basin rim (Figs 5.1A and B).

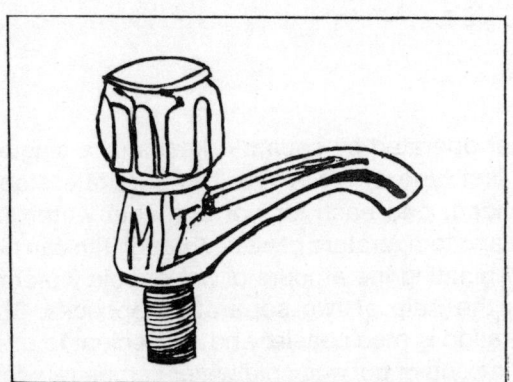

Fig. 5.1A *Pillar cock with vertical inlet for water connection.*

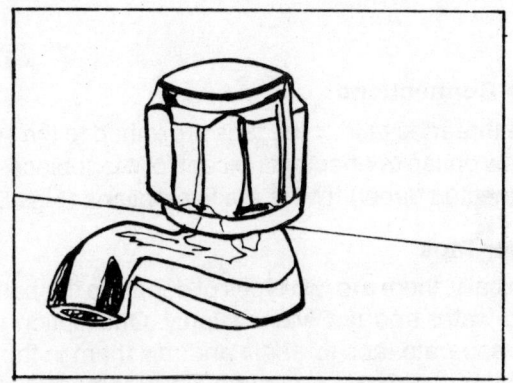

Fig. 5.1B *Pillar cock fixed on the basin rim.*

Bib Cocks

Bib cocks are the taps with horizontal inlets and are mounted on the wall (Figs 5.2A to C).

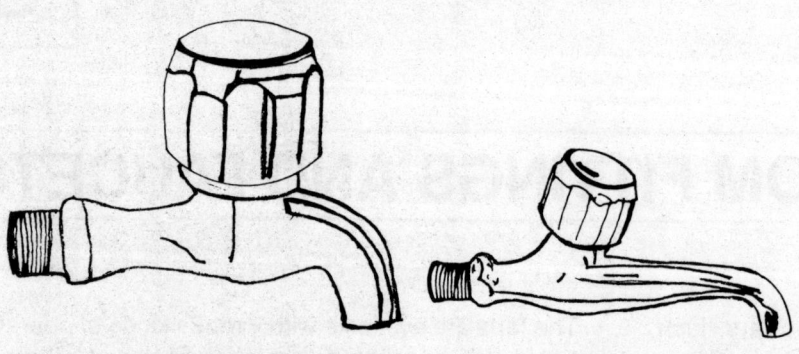

Fig. 5.2A *Standard bib cock.*

Fig. 5.2B *Long nose bib cock.*

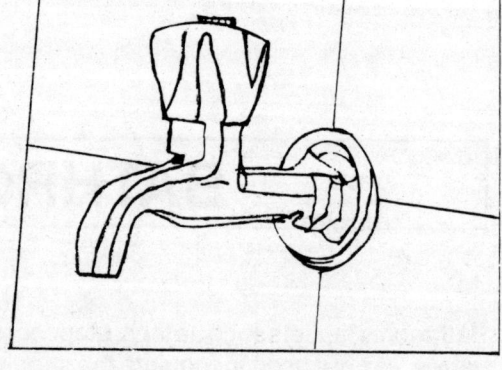

Fig. 5.2C *Wall-mounted bib cock.*

Stopcocks

Stopcocks are used to control the flow of water entering the taps. Stopcocks are fitted before each appliance and are used to regulate the quantity of hot/cold water in the mixer taps (Figs 5.3A to C).

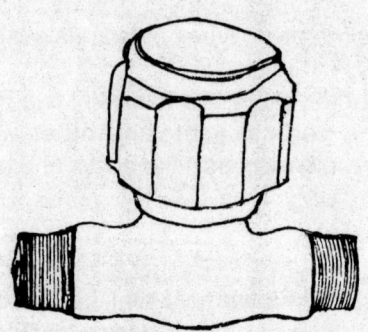

Fig. 5.3A *Standard stop cock with inlet and outlet.*

Fig. 5.3B *Standard concealed stop cock.*

Fig. 5.3C *Standard stop cock with inlet and outlet at an angle.*

Tap Connections

The threaded tails of the taps are joined to the supply with screw on tap connectors (except for monoblocs, which are connected direct). There are five options (Figs 5.4A to E).

Mixer Taps

Basically, there are two types of supply to the bathrooms—cold water and hot water supply. One option is to have two separate taps for them and mix them in the bucket or bathtub to achieve a desired water temperature. But the use of mixer taps can make mixing trouble free, by mixing the hot and cold water in the tap itself. Mixer taps are either operated by separate knobs or a single lever. For the first type of mixer tap, two separate stopcocks are provided, one each for hot and cold water supply. The flow and temperature of water through tap can be achieved by regulating the amount of hot or cold water as desired with the help of two separate stopcocks. Single lever operation is much easier and convenient to use. The flow and mixing of hot and cold water can be regulated by just changing the direction of a single lever, which is very easy to operate. Mixer taps are either wall mixers or basins mixer.

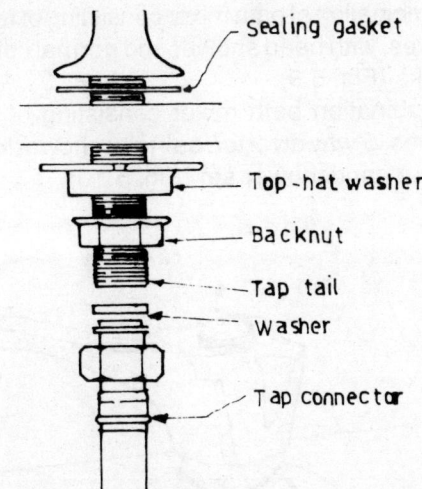

Fig. 5.4A *A tap connector with fibre sealing washer.*

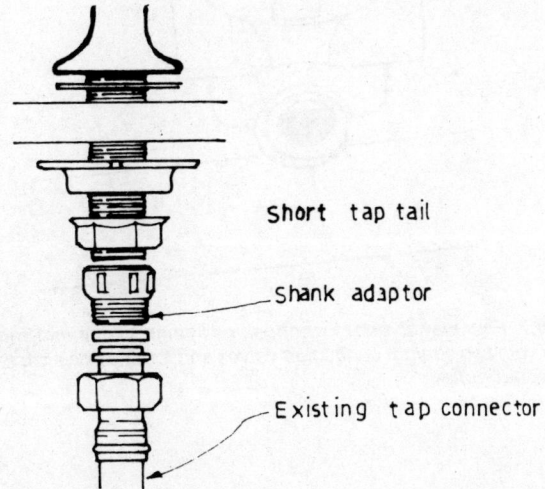

Fig. 5.4B *Shank/sink adaptors used to bridge small gaps.*

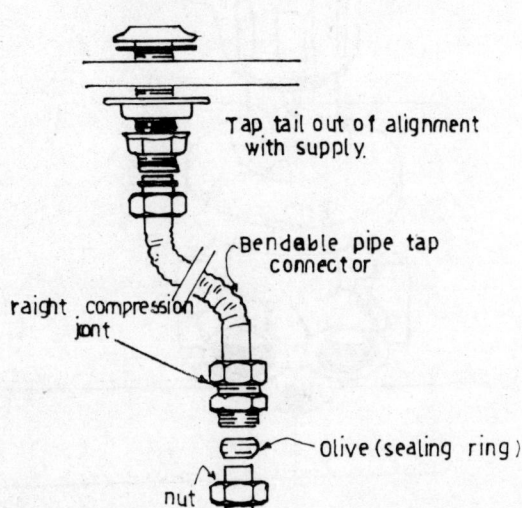

Fig. 5.4C *Bendable pipe tap connector kits span awkward gaps.*

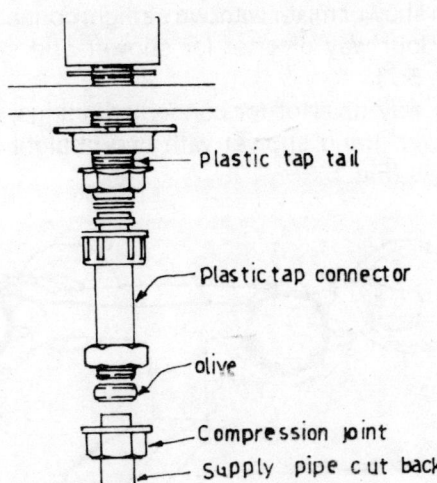

Fig. 5.4D *Plastic connectors used with some plastic taps.*

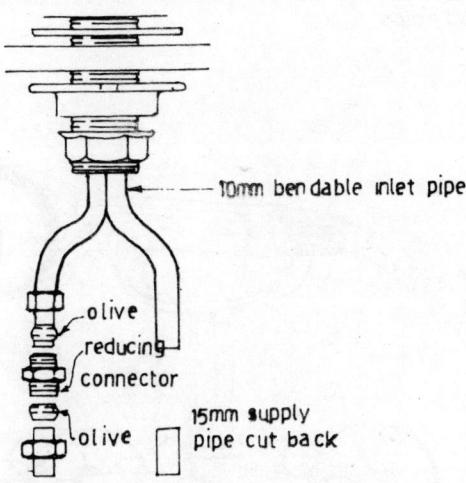

Fig. 5.4E *Reducing connectors joining 10 mm dia inlets to 15 mm supply pipes.*

Wall Mixers

Wall mixers are the mixer taps which are mounted on the wall and are mostly used in the bathing area. Mixer taps are available in different varieties and designs to choose from. Wall mixers are either operated by a single lever or two knobs—for hot and cold water.

Following are the varieties of wall mixers commonly available:

Wall mixers with separate knobs

1. Wall mixer with two knobs—one for hot water and one for cold water (Fig. 5.5).
2. Wall mixer with telephonic shower arrangement including shower rest, flexible tube and shower (Fig. 5.6).

3. Four-way diverter for concealed fittings—concealed bath shower mixer with two straight concealed valves and four way diverter for shower and spout outlet (Fig. 5.7).

4. Five way divertor for concealed fittings, concealed shower, hand shower with two straight concealed valves (Fig. 5.8).

5. Combination of bath mixer consisting of two concealed valves, with head shower and normal concealed stop cocks (Fig. 5.9).

6. Combination bath mixer consisting of 2 concealed valves, 5 way divertor, ball joint shower with arm, wall spout handshower set (Fig. 5.10).

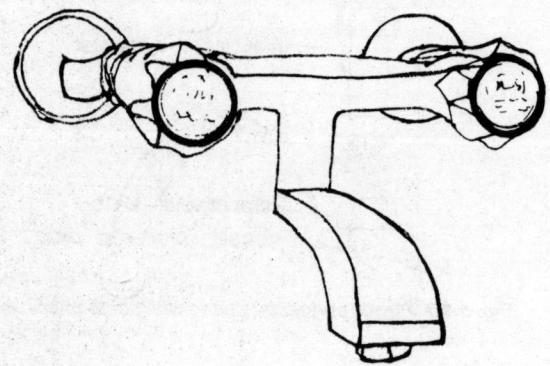

Fig. 5.5 *Wall mixer with two knobs one for hot water and one for cold water.*

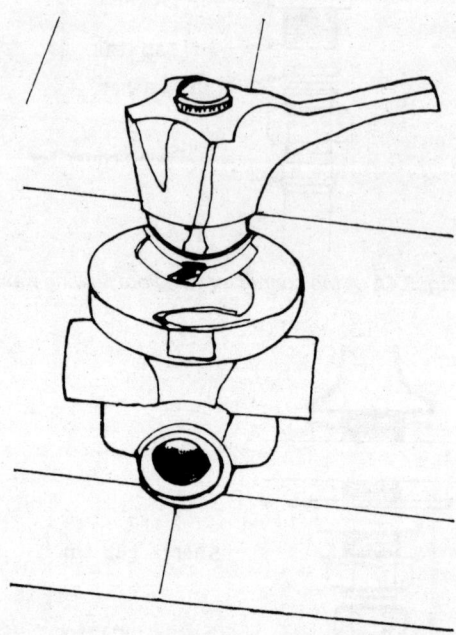

Fig. 5.7 *Four way divertor for concealed fittings, concealed bath shower mixer with two straight concealed valves and four way divertor for shower and spout outlet.*

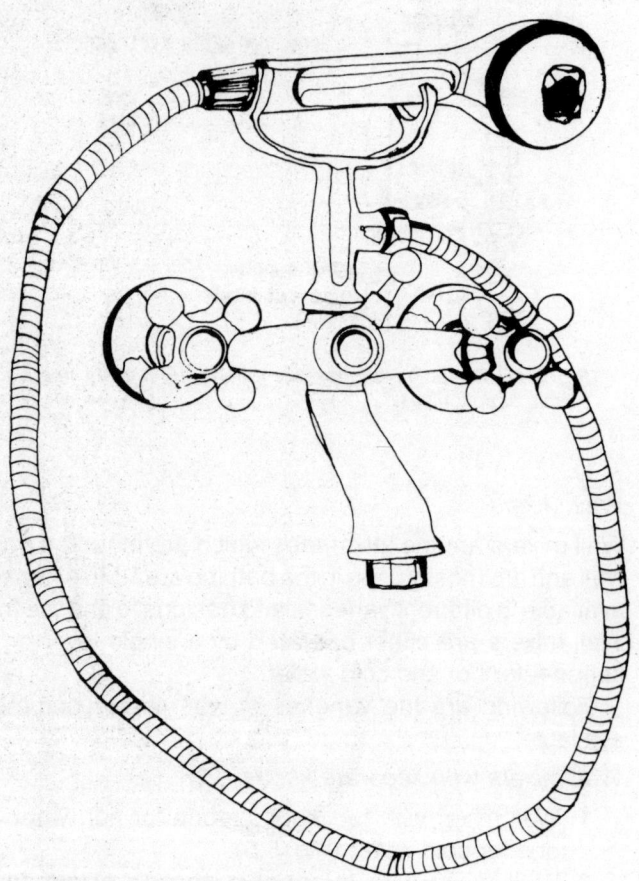

Fig. 5.6 *Wall mixer with telephonic shower arrangement including shower rest flexible tube and shower.*

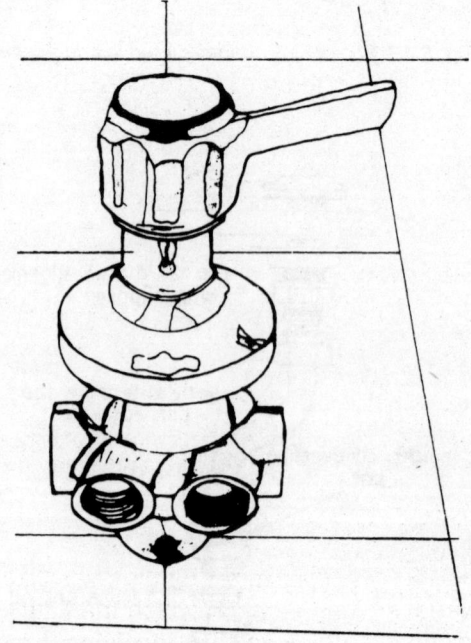

Fig. 5.8 *Five way divertor for concealed fittings, concealed shower, hand shower with two straight concealed valves.*

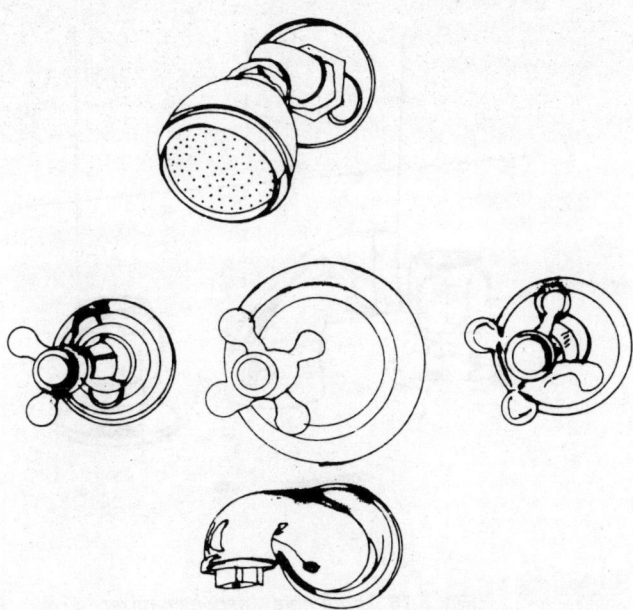

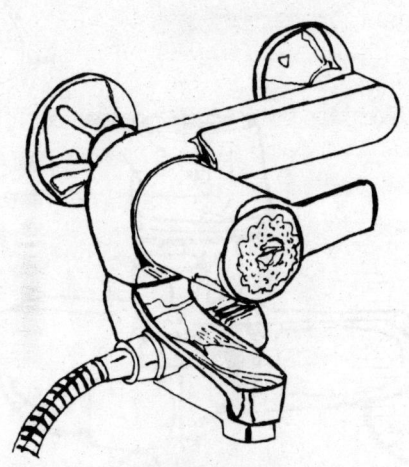

Fig. 5.11 *Single lever operation for bath as well as shower for hot as well as cold water.*

Fig. 5.9 *Combination of bath mixer consisting of two concealed valves, with head shower and normal concealed stop cocks.*

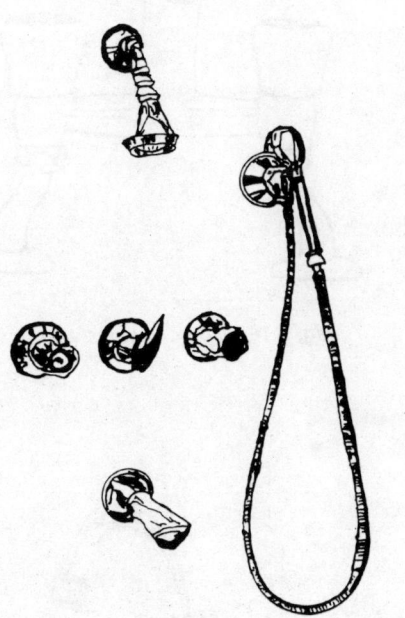

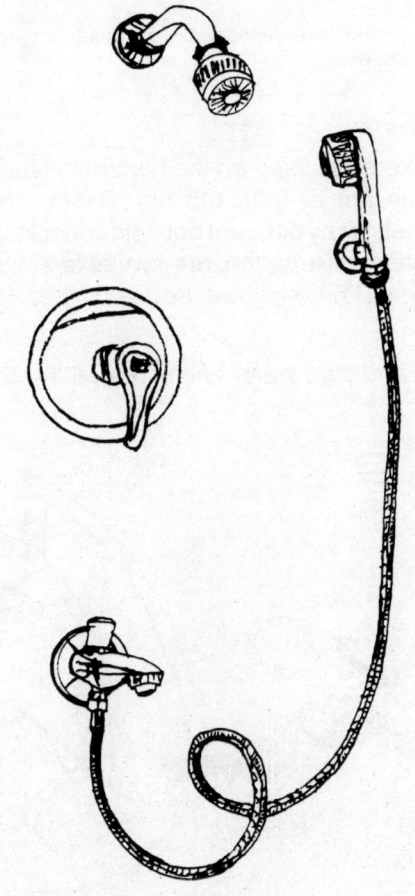

Fig. 5.10 *Combination mixer consisting of two concealed valves, five way divertor, ball joints shower with arm, wall spout and hand shower set.*

Single lever bath mixers Single lever mixers are available in various types—one in which the fittings are exposed, and another one in which all the fittings are concealed with only lever and spout open to view (Fig. 5.11, 5.12 and 5.13).

Fig. 5.12 *Single lever 4 ways fitting: Concealed divertor for bath and shower arrangement an obstructive fitting for regulating the shower/spout outlets.*

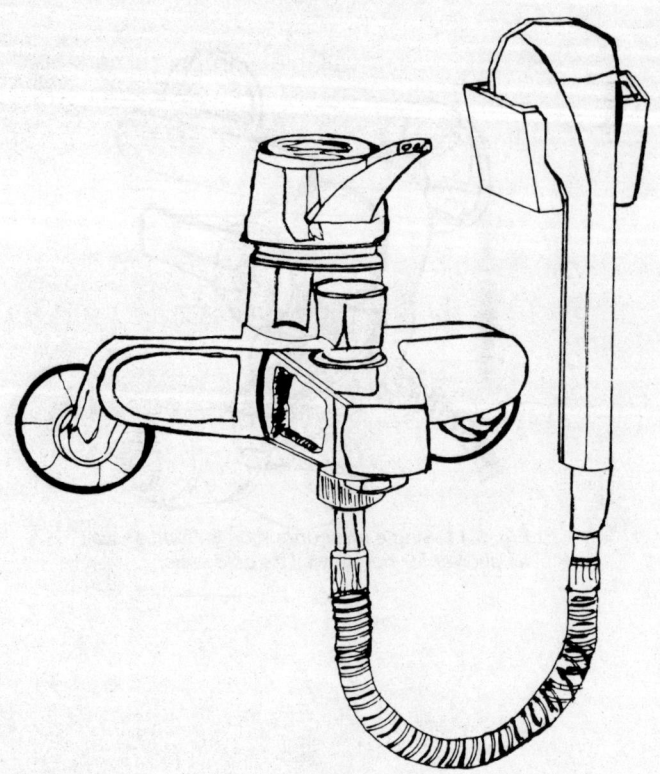

Fig. 5.13 *Single lever exposed type mixer for bath/shower with telephonic shower.*

Basin Mixers

Basin mixers are fixed on the horizontal surface either on washbasin rim or bath tub rim. Basin mixers are also operated either by different hot/cold water knobs or a single lever—basin mixers, too, are available in various shapes and designs. Following are the basic types of basin mixers available.

With hot and cold water knobs (Fig. 5.14, 5.15, 5.16 and 5.17).

Fig. 5.15 *Three hole washbasin mixer.*

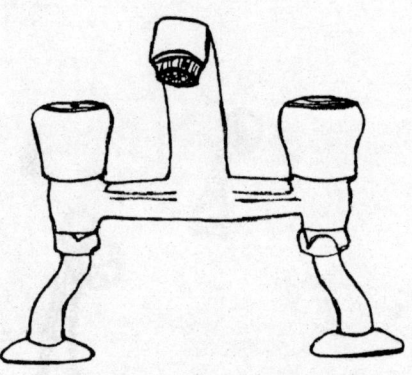

Fig. 5.16 *Two hole washbasin mixer.*

Fig. 5.14 *Wash basin mixer with hot and cold water knobs.*

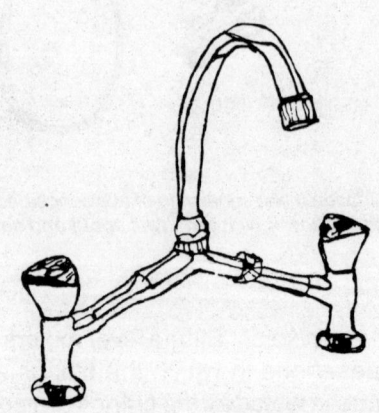

Fig. 5.17 *Two hole washbasin mixer, pillar mounted adjustable (bridge types).*

Single lever basin mixer It is connected to the hot water and cold water supply. A change in the direction of the lever gives hot water or cold water, or any temperature in between. The same lever controls the water flow, too. More water flow, when it moves up and less water when it moves down (Figs 5.18A and B).

Bidet Mixers

Bidet mixers are usually available with pop-up drainage system. These are of two types (Figs 5.19 and 5.20).

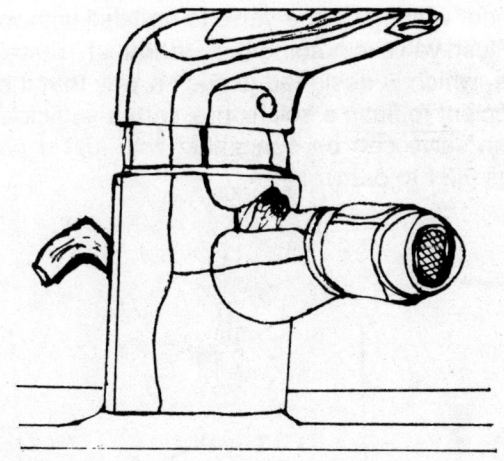

Fig. 5.19 *Single lever bidet mixer with pop up waste.*

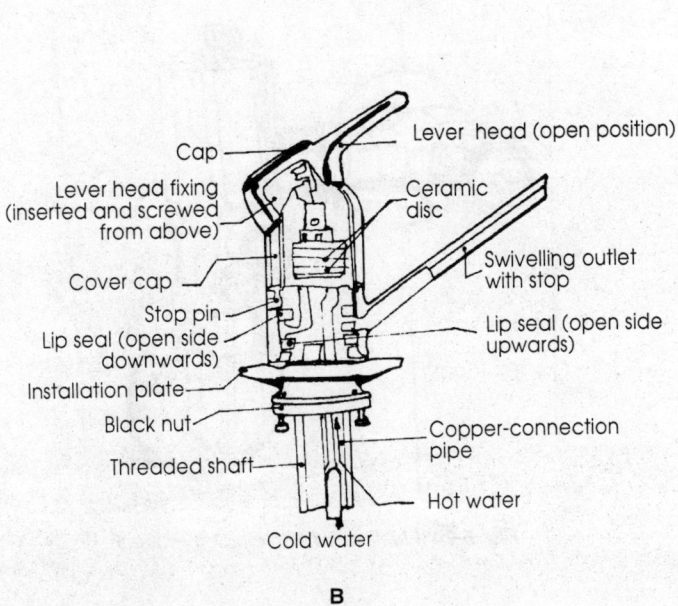

A

B

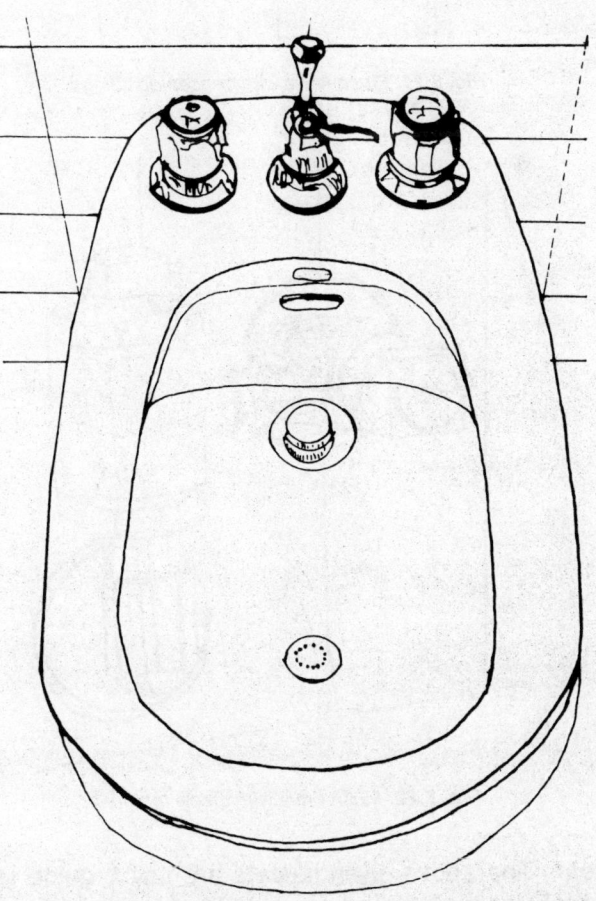

Figs 5.18A and B *Sketches show single lever wash basin mixer. It is connected to the hot water and cold water supply outlets. A change in the direction of the lever gives hot or cold water or any adjusting the flow of supply to moderate the temperature. The same lever controls the water flow. By moving the lever upwards there is more flow and moving it downwards reduces the flow and absolutely downward stops water flow.*

Fig. 5.20 *Four hole bidet mixer complete with shower and pop-up waste.*

FLUSH VALVES

Flush valves are used for flushing the water closets or urinals. They are the replacement of flushing cistern and have various advantages—it takes less space as compared to the flushing cistern as there is only a small valve. The flush valve can be flushed continuously one after another without waiting for the valve to be filled with water.

Flush valve is nothing but a length of 150 mm diameter pipe, which is designed in such a way that it holds water sufficient to flush a soil fixture, with a sufficient pressure. Flush valve can be concealed with just a proper fitting (Figs 5.21 to 5.24).

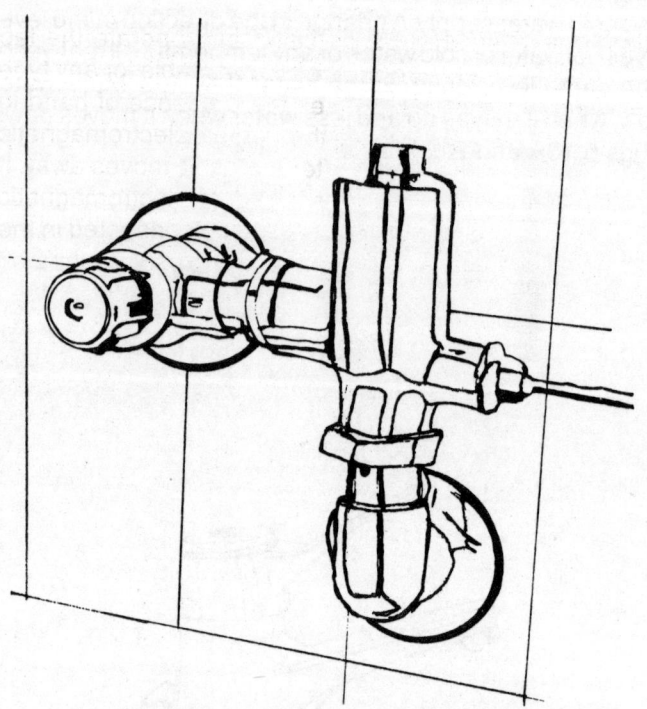

Fig. 5.23 *Flush valve with elbow with side push.*

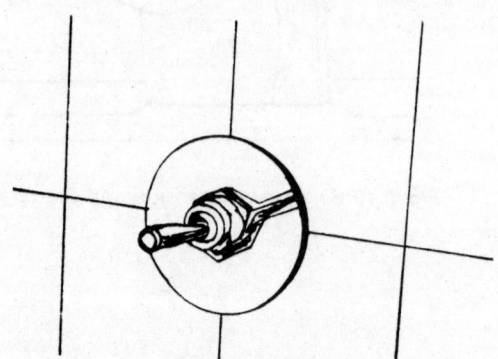

Fig. 5.21 *Flush valve with concealed fittings.*

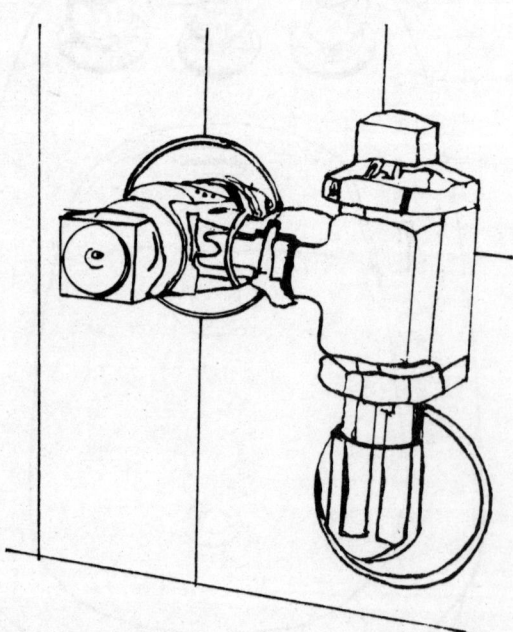

Fig. 5.22 *Flush valve with elbow top push.*

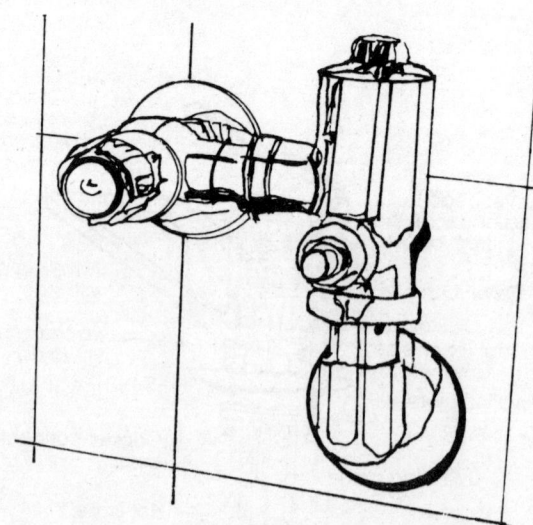

Fig. 5.24 *Flush valve with elbow front push.*

CONSERVING WATER

Imagine how much water is wasted due to people leaving the tap running in offices, hotels, factories, railway stations, airports, and other public places and hotels. Public awareness is definitely the solution, but it is very difficult to achieve it in near future. Another solution is to use the water conserving faucets having automatic controls.

Note The above given faucets are just a guide to the basic types available in the market. The faucets are available in various design and colours under different tradenames.

Automatic Control Taps

Auto tap is designed to save water and operate the tap automatically whenever required. It is suitable for any type of existing tap. Auto tap senses the presence of hand in front of tap and allows water through an electromagnetic valves. Water stops flowing after the hand moves away. It consists of a control unit, sensor and an electromagnetic valve. The electromagnetic valve is to be connected in the pipeline supplying water to the tap. Sensors are clamped above the tap.

Auto Stop Taps

Auto stop taps shut off automatically within 30 seconds (or the preset timings).

Electronic Flushing Unit-I

Electronic flushing unit is an automatic flush unit, which helps conserve water used in public toilets. In a conventional automatic system, water keeps on flowing regularly at fixed intervals, irrespective of whether the toilet is in use or not. The autoflush senses the presence of a person in front of the urinal or WC and supplies water through electromagnetic valve for a preset time, so that it is flowing only when required.

BATHROOM ACCESSORIES

At the initial designing stage, a little thought is given to the design of bathroom accessories. Whatever the item, from nailbrush to cupboard hinge, it should serve its purpose as functionally as possible, look as attractive as its function will allow, and be operative at the most convenient point. Choose accessories to match the style of your bathroom. Amongst accessories are towel rails, toilet roll holders, toothbrush holders, soap cases, electronic hand dryer, and mirrors.

They are available in plastic, chinaware, chromed metal or symet (synthetic metal) and are available in a wide range to choose from.

TOWEL RAILS

There should be towel rails near the shower, the bath and the washbasin. Though, sometimes, if the sanitary ware is adjacent, one towel rail can double for two uses. Towel rails come in the form of fixed rods, hinged brackets or rings, singly or in multiples, in metal, plastic, and timber (Figs 6.1A to 6.1F).

Rings are only suited to hand towels because larger towels will not have sufficient spread to dry out between use. Often, towel rails too will be a part of matched suit and come in a range of sizes to suit most requirement. Place

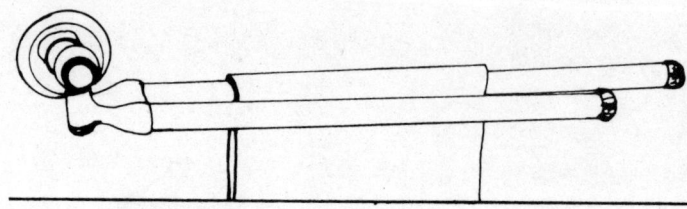

Fig. 6.1.C *Double towel rail (swivel).*

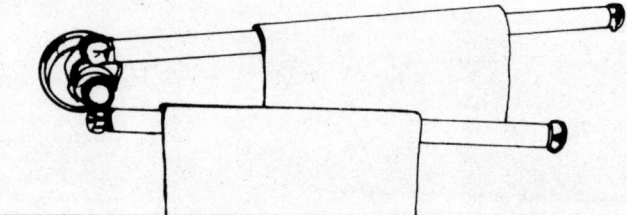

Fig. 6.1D *Swing arm towel rail.*

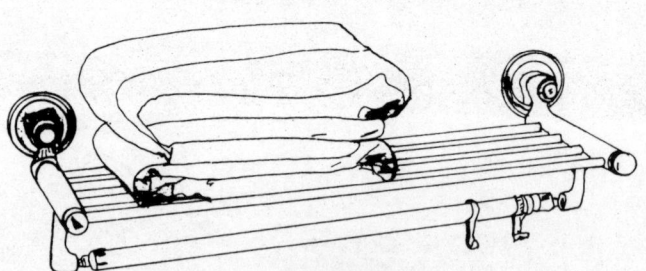

Fig. 6.1E *Towel bracket.*

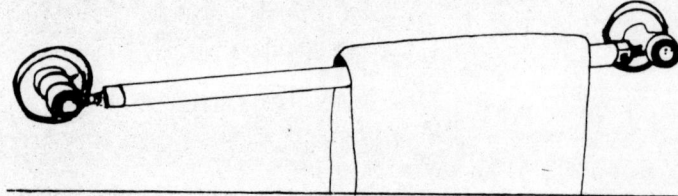

Fig. 6.1A *Single towel rail.*

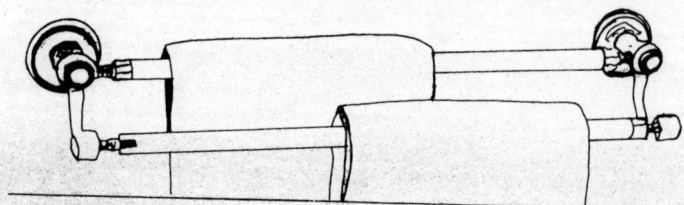

Fig. 6.1B *Twin towel rail.*

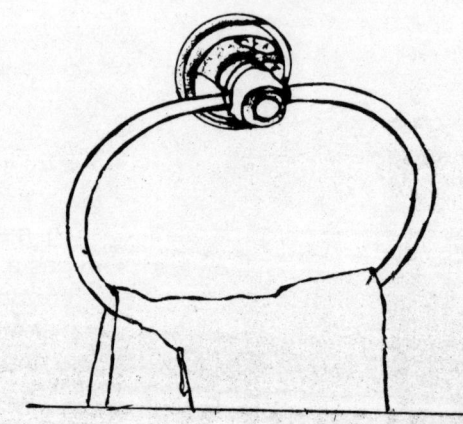

Fig. 6.1F *Towel ring.*

rails within safe reach of shower or tub. Some rails placed lower on wall will encourage children to be tidy.

TOILET PAPER HOLDERS

Toilet paper holders are more or less similar in styles, but their positioning is important. To one side of the lavatory and at low level is the best, not behind it and way above your head.

Holders can be of box type for interleaved tissue or designed to hold a roll of paper, provided they allow the roll to swing around unhindered. The latter are best recessed into the wall, with a spring-loaded rod (Figs 6.2A to C). Be sure to place the holder where it can be comfortably reached by all users, at optimum, this would be directly in front of the appliance, but when this is not possible to the right is better than the left, since the majority of people are right handed.

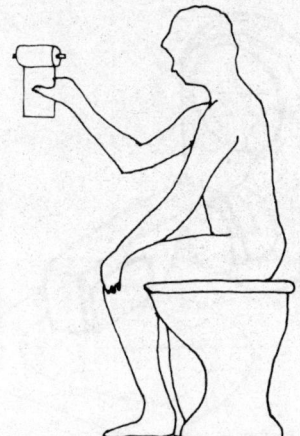

Fig. 6.2C *Position/location of toilet paper holder in bathroom.*

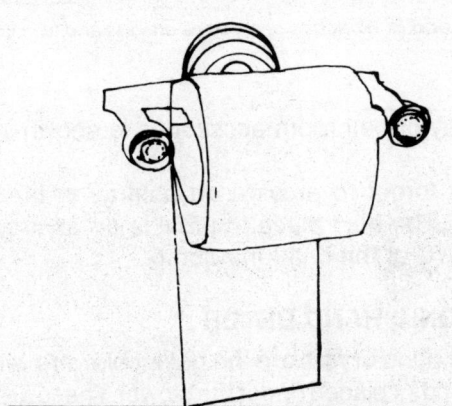

Fig. 6.2A *Toilet paper roll holder.*

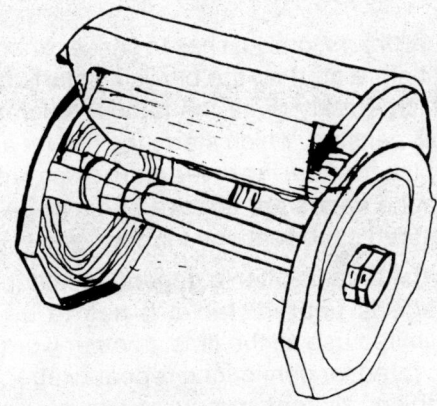

Fig. 6.2B *Toilet paper holder with cutter.*

SOAP CASES

Soap cases can be recessed in the wall or projecting one. Whatever type of soap dish you are using, make sure that it is drained properly as the soap becomes a soggy mess most of the times, and the consumption rate is doubled. Holding all plastic or metal trays which stretch from one side to the other does the job fairly well, and is useful for sponges brushes, etc. at the same time. Most satisfactory of all are the magnetic soap dishes and rubber suction

pads. Rubber pads can be attached to a vertical surface— the wall beside the bath or shower, and hold the soap vertically instead of horizontally, water can actually drain according to the natural laws of gravity (Figs 6.3A to C).

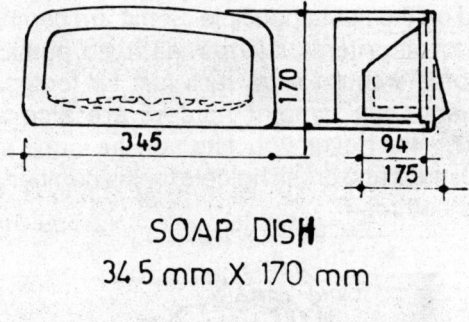

SOAP DISH
34.5 mm X 170 mm

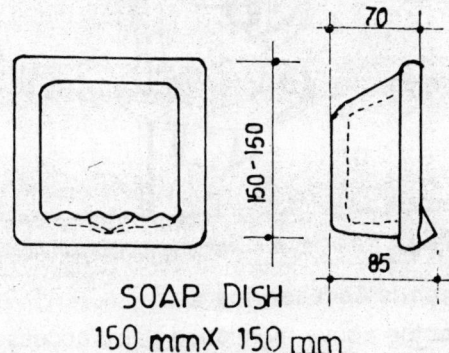

SOAP DISH
150 mmX 150 mm

Fig. 6.3A *Various types of soap dishes in Chinaware.*

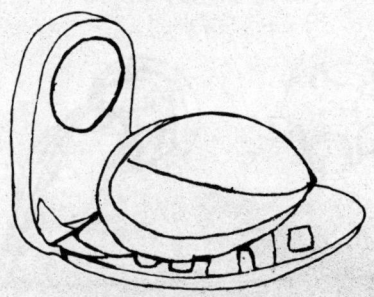

Fig. 6.3B *CP soap dish.*

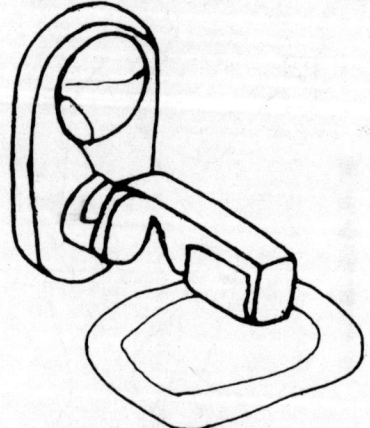

Fig. 6.3C CP magnetic soap dish.

TOOTHBRUSH HOLDERS

A toothbrush which is constantly returned wet to a cupboard is doomed to a very short life. Thus, it is preferable that a suitable rack is located in the open, and has just a space for the toothpaste to be kept along. Toothbrush racks are all designed on the same principle. Some are combined with a hold all for other items, such as nail brush, pumic, etc. and others come with an extra rack just for toothpaste and toothbrush. Only tumbler holders are also available (Fig. 6.4). Whichever you choose, be sure there are sufficient mugs and brush holders for the number of people using the bathroom.

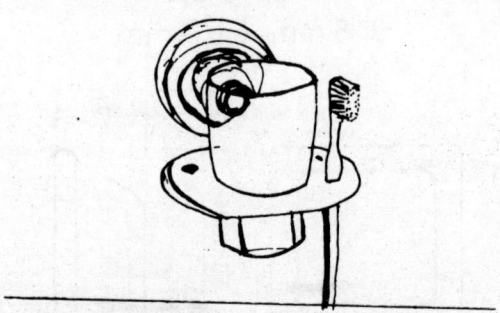

Fig. 6.4 CP tumbler holder and toothbrush holder.

Miscellaneous Accessories

Apart from the above mentioned basic accessories, few more things are also required in the bathroom like, bath robe hook, pegs for clothes, etc. these pegs should be provided in the dry area of the bathroom (Figs 6.5 to 6.6B).

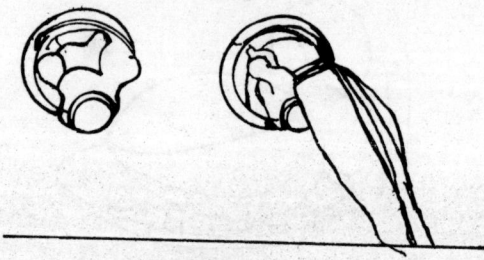

Fig. 6.5 CP bath robe hook.

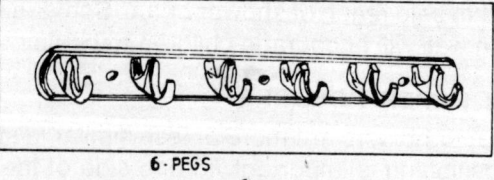

6 - PEGS

A

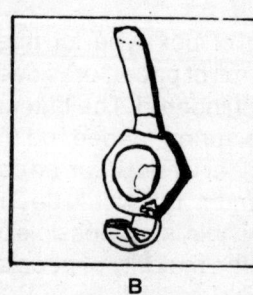

B

Figs 6.6A and B Sketches of six pegs bracket and twin peg bracket.

An away of bathroom accssories is shown in Figs 6.7 and 6.8.

Do not forget to provide an ashtray in the smoker's bathroom. The best place to provide an ashtray is either near the WC or the bathtub or both.

ELECTRONIC HAND DRYER

Electronic hand dryers are the hot air blowers, which dries the wet hands if placed under the blower. These are normally used in hotels, hospitals, etc. for a hygienic replacement of the towels. These dryers are either manually operated or automatic.

In manual dryers, one just has to shake excess water off the hands before starting the blower. It starts blowing out hot air, which quickly dries the hands. Automatic dryers have in-built censors, which starts the blower as soon the hands are placed under the dryer, and it stops automatically as soon as the hands are removed from there.

The required specifications for any hand dryer are:

(i) As the dryer is often a general area equipment, its usage is unpredictable. It has to be ready for unlimited use all the time. In other words, it has to be "rated for continuous repeat usage".

(ii) Ambient air temperature varies considerably for winter to summer. Thus, it is very important to have "summer winter control".

(iii) The standard accepted rating for a hand dryer is "the time required to dry a pair of hand up to wrists is approximately half a minute".

MIRRORS

Bathrooms are only half alive without the addition of mirrors. Mirrors add a touch of glamour to a bathroom and give a sense of space. Their use vary from the large sheets of glass covering the whole wall to the small mirror above a

Plate 1

Washbasins

Various types of washbasins and their applications with counter tops; in various shapes, colours and usages.

WASHBASIN RHAPSODY
A stylish and spacious contertop oval washbasin with twin soap trays.
Size: 63 cms × 50 cms.

WASHBASIN TIFFANY
A wide oval basin with luxurious wash space and twin ribbed soap trays on the side.
Size: 51 cms × 38 cms.

WASHBASIN OVAL
An undercut oval basin that is both modern and simple in design.
Size: 55 cms × 40 cms.

WASHBASIN STARLET
An aesthetically appealing, compact, countertop round basin with twin soap space with decorative pleats. May be installed singly or in series.
Size: 38 cms dia & 46 cms dia.

TWIN WASHBASIN

WASHBASIN CORNER
A compact design that fits snugly into a corner, ideal where space is at a premium. Fixed on concealed Z type bracket or screwed to the wall.
Size: 45 cms × 39 cms.

WASHBASIN WINDSOR
A stylish yet practical basin that is ideal for space economy bathrooms. Screwed to the wall or fixed on brackets.
Size: 45 cms × 33 cms.

WASHBASIN COMPACT
A traditionally styled basin that lends a tasteful, uncluttered look to small bathrooms. Fixed on brackets or screwed to the wall.
Size: 45 cms × 30 cms.

Plate 2
Washbasins and Pedestals

Various types of washbasins with pedestals. The pedestals cover the CP bottle traps and angle valves for elegance and beauty to enhance the aesthetics of the decor.

WASH BASINS & PEDESTALS

BATHROOM DECOR
Washbasin with pedestal in a modern bathroom setting.

WASHBASIN STANDARD
A conventional design with back skirting that brings out a fresh, modern appeal in economy bathrooms. Can be fixed on rag bolts or brackets, with or without pedestal.
Size: 55 cms × 40 cms & 63 cms × 45 cms.

WASHBASIN SOLA
A stylish all bowl design with splash-resistant rim, that lends character to the elite bathrooms. Suitable for fixing on rag bolts or brackets, with or without pedestal.
Size: 58 cms × 45 cms & 51 cms × 40 cms.

WASHBASIN SAVOY
A washbasin inspired by a sense of romantic curves, decorative pleats, graceful shape and gracious styling. Fixed on rag bolts, with or without pedestal.
Size: 55 cms × 40 cms.

WASHBASIN NOVA
An opulent design for spacious bathrooms, suitable for fixing on brackets or rag bolts, with or without pedestal.
Size: 63 cms × 51 cms.

WASHBASIN VIKING
An elegant model with warm, classic contours, featuring a chic rectangular bowl, that will blend with any design. Suitable for fixing on rag bolts or brackets, with or without pedestal.
Size: 55 cms × 40 cms.

WASHBASIN MICHELANGELO
A beautiful, functional design, that neatly fits into any bathroom, yet makes for an individual presence. Fixed on rag bolts or brackets, with or without pedestal.
Size: 58 cms × 43 cms.

WASHBASIN SOLARA
A distinctive basin with bold, clear lines, having generous washing space and a flat shelf area across the back. Fixed on rag bolts, with or without pedestal.
Size: 55 cms × 40 cms.

WASHBASIN SUPER CONSTELLATION
An attractive basin with a unique styling, having a well-emphasized rim. Can be fixed on brackets or rag bolts, with or without pedestal.
Size: 52 cms × 42 cms.

WATER CLOSETS

EWC UNIVERSAL
A closet with twin functionality. Can be used as a squatting WC or a European WC. Used with cistern. Available in P or S trap.
Height: 81 cms.   Width: 50 cms.
Projection: 80 cms.

EWC SAVOY
A luxurious designer suite that blends classic elegance with style, with a central pull lever on the cistern. Flushes efficiently with 6.5 ltrs of water.
Height: 80 cms.   Width: 37 cms.
Projection: 66 cms.

EWC SYPHONIC
An elegant water closet incorporating a double trap syphonic flushing action for quiet functioning. The cistern with bottom inlet adds to the aesthetics. Available in P or S trap.
Height: 80 cms.   Width: 51 cms.
Projection: 75 cms.

EWC SOLARA
A sleek, designer suite, with handsome, geometric contours and a stylish side press lever for the cistern, blending looks and comfort. Flushes efficiently with 6.5 ltrs of water.
Height: 81 cms.   Width: 40 cms.
Projection: 72 cms.

EWC MICHELANGELO
A sophisticated, futuristic EWC, with a sleek, tall cistern, specially for the bathroom with distinction. Pneumatic flushing system with push top button, flushes efficiently with 7.5 ltrs of water.
Height: 80 cms.   Width: 35 cms.
Projection: 67 cms.

EWC CONSTELLATION
A slender, curvaceous design with trendy, elegant looks. Unique water saving mechanism that flushes in just 5litrs of water. Special fitting with pull lever on top. Available in P or S trap.
Height: 80 cms.   Width: 36 cms.
Projection: 72.5 cms.

EWC STANDARD
A neatly designed inexpensive, traditional model, with an efficient wash down bowl, used with low level cistern. Available in P or S trap, with vent options.
Height: 89.5 cms.   Width: 50 cms.
Projection: 78 cms.

EWC SUPER CONSTELLATION
(Registered Design)
Gracefully contoured with rich, sculptured looks, that lends an exquisite charm. A pneumatic flush system with push button on top, that flushes with just 3-5 ltrs. of water. Available in P or S trap.
Height: 80 cms.   Width: 38 cms.
Projection: 70 cms.

Plate 4
Bidets and Squatting Pans
Various types of bidets available from floor mounted to wall mounted with various types of CP and chrome fittings, etc. The squatting type pans ranging from ordinary Indian pan to Orissa pan.

BIDETS

BIDET CONSTELLATION
A slender, trendy & elegant design to complement the CONSTELLATION suite.
Height: 41 cms.   Width: 36.5 cms.
Projection: 60 cms.

BIDET STANDARD
A luxury bidet for opulent bathrooms, with the popular ascending spray and the added comfort of warm water heated rim.
Height: 38.5 cms.   Width: 34.5 cms.
Projection: 60 cms.

BIDET WALL MOUNTING CONSTELLATION
Delicate and appealing in design, a wall mounting bidet specially designed to match wall mounted closets. Fixed on rag bolts.
Height: 38 cms.   Width: 37.5 cms.
Projection: 62 cms.

BIDET SAVOY
A luxurious bidet blending elegance with style, to match the SAVOY suite, with over the edge water supply.
Height: 43 cms.   Width: 39 cms.
Projection: 60 cms.

BIDET SOLARA
A sleek, designer bidet with geometric contours, to perfectly match the SOLARA suite.
Height: 43.5 cms.   Width: 39.5 cms.
Projection: 63.5 cms.

SQUATTING PANS

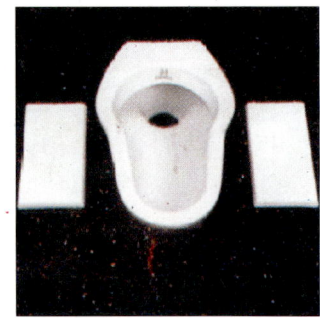

INDIAN PAN
A specially designed squatting pan with back projection to facilitate firm placement.
Size: 45 cms & 51 cms.

INDIAN PAN
The most common and widespread squatting pan, suitable for typical economy bathrooms. Available with back inlet.
Size: 51 cms & 58 cms.

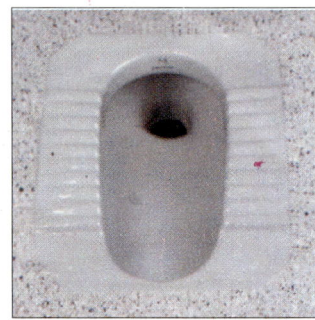

ORISSA PAN
Deluxe squatting pan with box rim and chamfered corners for a contemporary look. Suitable with low level cistern.
Size: 58 cms × 44 cms.

Plate 5
Bathtubs
Various types of bathtubs available with their usages with shower cabins and bathing pools.

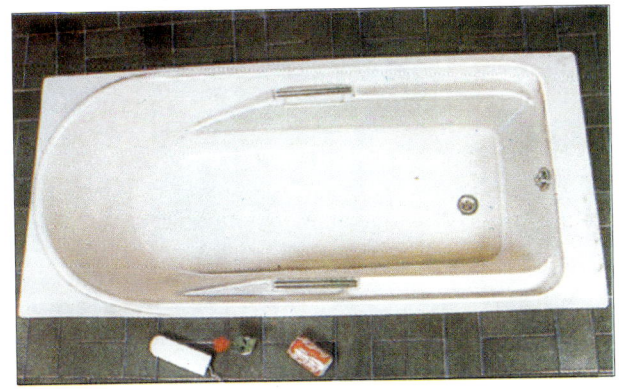

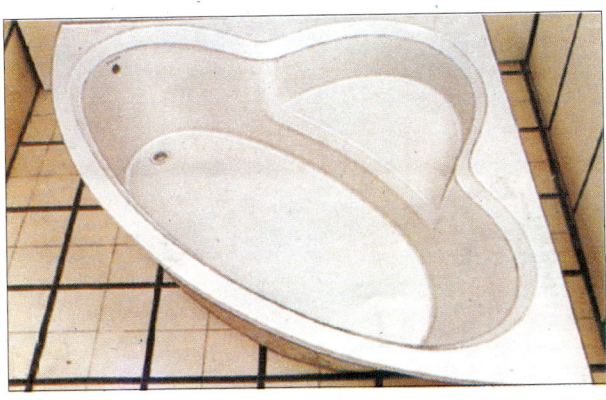

Plate 6
Bathroom CP Fixtures

• Various CP fixtures with CP chrome/powder coated/acrylic knobs with half-turn fittings.
• Quarter-turn fittings and Queen's design CP powder coated bathroom fittings.
• Single lever 4–5-way CP fittings.

Plate 7
Combination of Materials and Colours
Use of various combinations of materials and colours in bathrooms.

Plate 8
Tiles and borders
Use of coloured tiles, borders, corners and mouldings.

Plate 9

Colours and Materials

Colours and use of ceramic tiles and marble stone, wood, carpet and synthetic flooring.

Colour wheel: *Colours that are adjacent on the wheel are complementary, those opposite to one another are contrasts usually used for accents.*

Designing with Colour

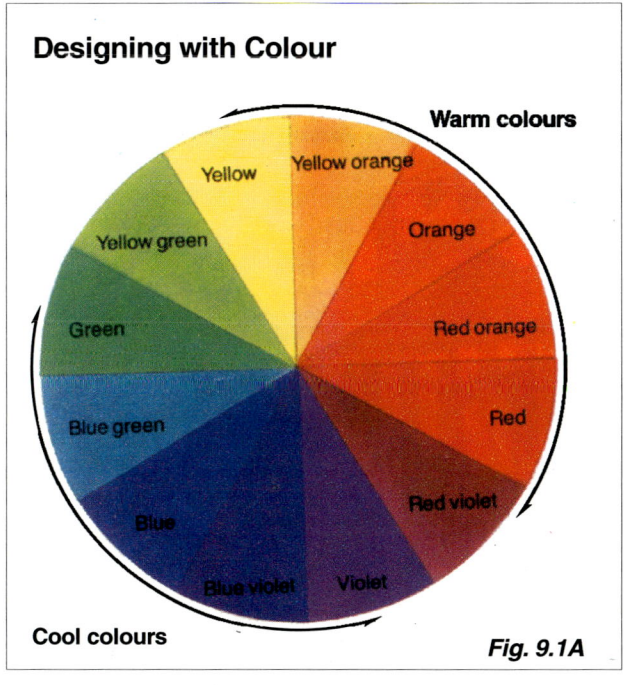

Fig. 9.1A

Fig. 9.1B

Use of ceramic tiles in wall dado, flooring and steps in bathrooms.

Use of marble stone flooring.   *Use of carpet flooring.*

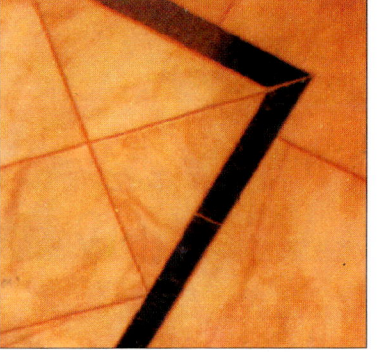

Use of synthetic, vinyl rubber or polyurethane resilients in bathrooms.

Use of wooden flooring.

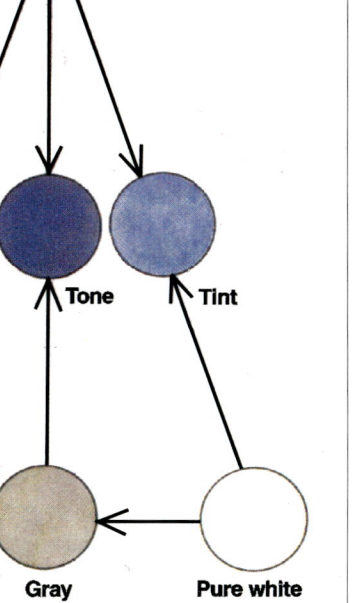
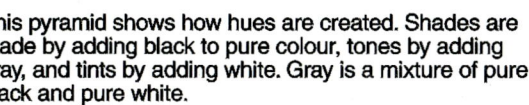

This pyramid shows how hues are created. Shades are made by adding black to pure colour, tones by adding gray, and tints by adding white. Gray is a mixture of pure black and pure white.

Plate 10
Materials and Applications
Use of wall paper, marble stone, glass blocks, ceramic tiles and wooden flooring.

Application of wall paper.

Use of marble stone on dados.

Application of glass blocks in walls.

Application of wooden flooring and taps.

Use of ceramic tiles.

Plate 11
Materials and Applications
Use of ceramic tiles, wood, lamination, polymers and marble stone in washbasin counters.

Use of ceramic tiles in washbasin counters.

Use of wood for washbasin counter.

Use of POP mouldings, cornices, pillars, etc.

Use of lamination in washbasin counters.

Use of polymers in washbasin inbuilt casted counters.

Use of marble stone in washbasin counters.

Plate 12
Glazed Tiles
Various types of glazed tiles available with colour circle for selection of style and colour.

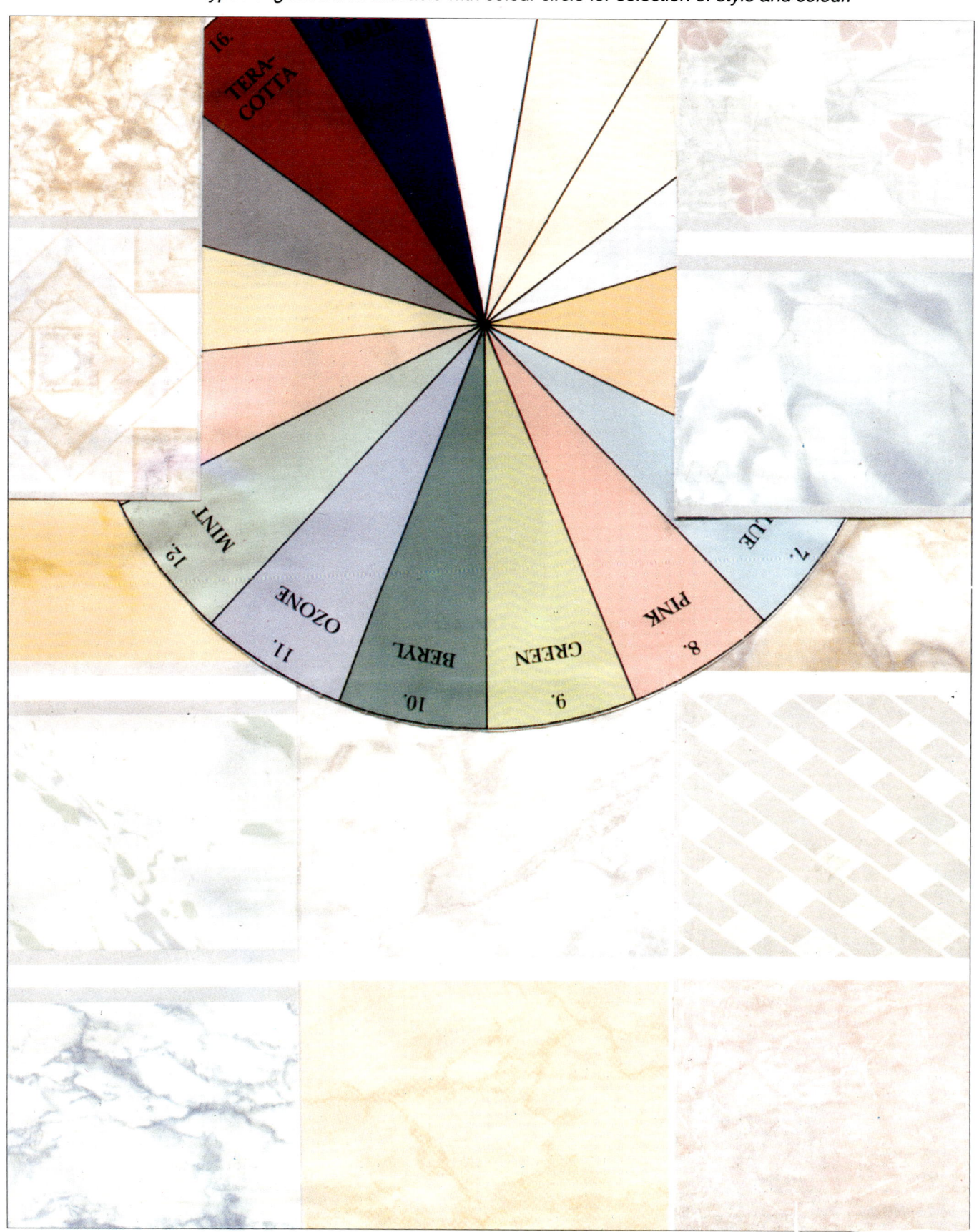

Plate 13
Glazed Tiles
Various designs and sizes of glazed tiles ranging from glazed, matt to textured finishing.

Plate 14
Granite Stone
Various types and colours of granite stone and its application as floor and wall tiles and kitchen and bathroom counters with various types of mouldings. Granite is a very hard stone.

1. Cobra Black   2. Jhansi Red   3. Rani Pink   4. Raniwara
5. Mokalsar Green 6. Hassan Green 7. Sadarli Gray 8. Z Black
9. Sira Gray 10. Multi Colour 11. Bangalore Red 12. Rosy Pink
13. Ruby Red 14. Cheema Pink 15. Jhunjhunu Red 16. Paradiso

Plate 15
Italian Stone

Various types of Italian stones available in India with their names. It is mostly applied in flooring and dados. It can also have various moulding battens similar to granite stone. It is a soft finished stone.

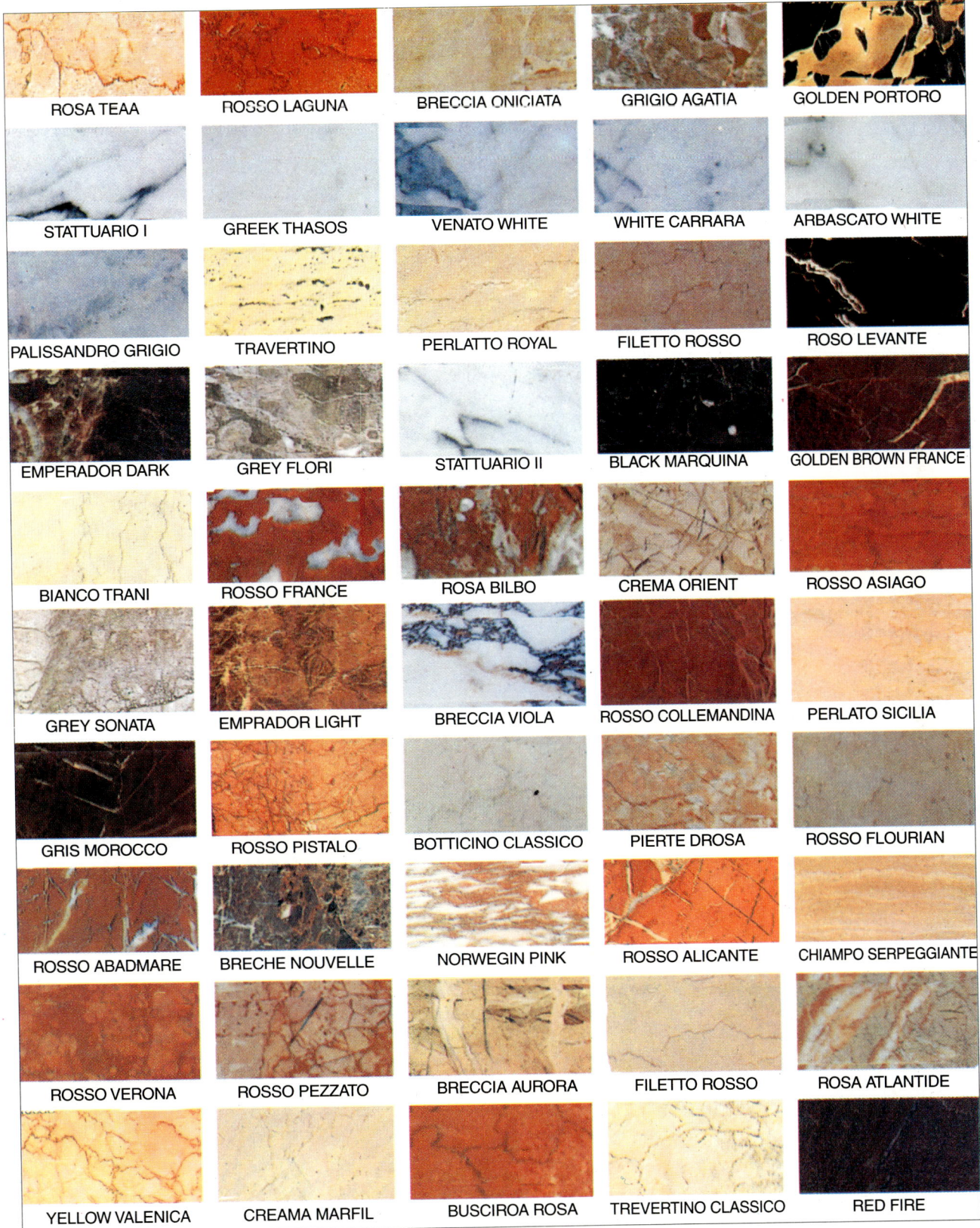

ROSA TEAA	ROSSO LAGUNA	BRECCIA ONICIATA	GRIGIO AGATIA	GOLDEN PORTORO
STATTUARIO I	GREEK THASOS	VENATO WHITE	WHITE CARRARA	ARBASCATO WHITE
PALISSANDRO GRIGIO	TRAVERTINO	PERLATTO ROYAL	FILETTO ROSSO	ROSO LEVANTE
EMPERADOR DARK	GREY FLORI	STATTUARIO II	BLACK MARQUINA	GOLDEN BROWN FRANCE
BIANCO TRANI	ROSSO FRANCE	ROSA BILBO	CREMA ORIENT	ROSSO ASIAGO
GREY SONATA	EMPRADOR LIGHT	BRECCIA VIOLA	ROSSO COLLEMANDINA	PERLATO SICILIA
GRIS MOROCCO	ROSSO PISTALO	BOTTICINO CLASSICO	PIERTE DROSA	ROSSO FLOURIAN
ROSSO ABADMARE	BRECHE NOUVELLE	NORWEGIN PINK	ROSSO ALICANTE	CHIAMPO SERPEGGIANTE
ROSSO VERONA	ROSSO PEZZATO	BRECCIA AURORA	FILETTO ROSSO	ROSA ATLANTIDE
YELLOW VALENICA	CREAMA MARFIL	BUSCIROA ROSA	TREVERTINO CLASSICO	RED FIRE

Plate 16
Use of Stained Glass and Marble Stone
Application of stained glass in various ways with innovative designs: Various floor patterns and staircase design with bathroom counter designs.

A farm house stair case done in stattuario

glasscapes

Bathroom counter top in Botticino

Integrated marble pattern for flooring

Brush & Paste
Stand Holder

417G Toilet Paper
Holder

414G Soap Dish

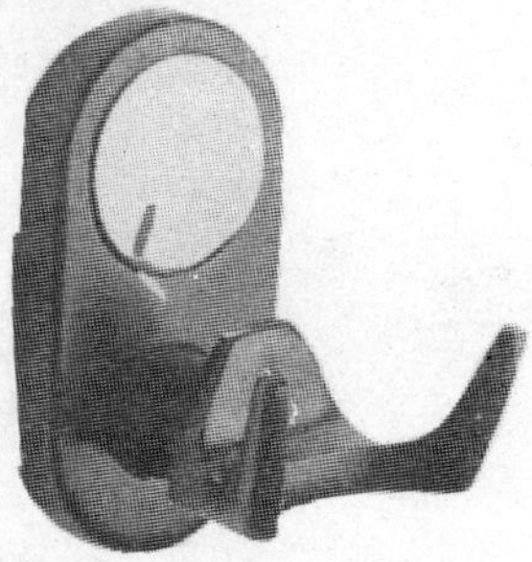

416G Robe Hook
Two way

420G Soap Dish
Magnetic

21119 Single Lever Wall Mixer

21213 Single Lever One Hole Bidet Mixer
with Pop Up waste system

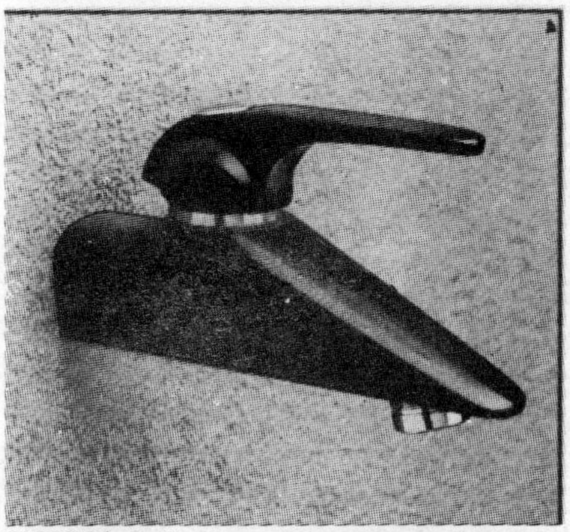

21037 Bib Cock

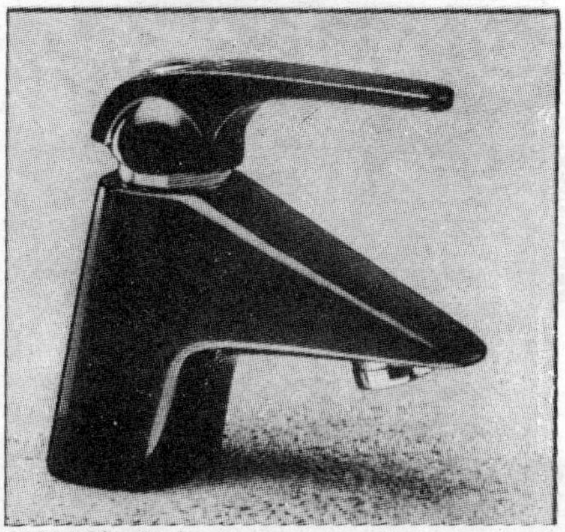

21001 Pillar Cock

21083 Concealed Stop Cock

21053 Angular Stop Cock

21463 Button Spout

21429 Bath Tub Spout

Fig. 6.7 An array of bathroom accessories.

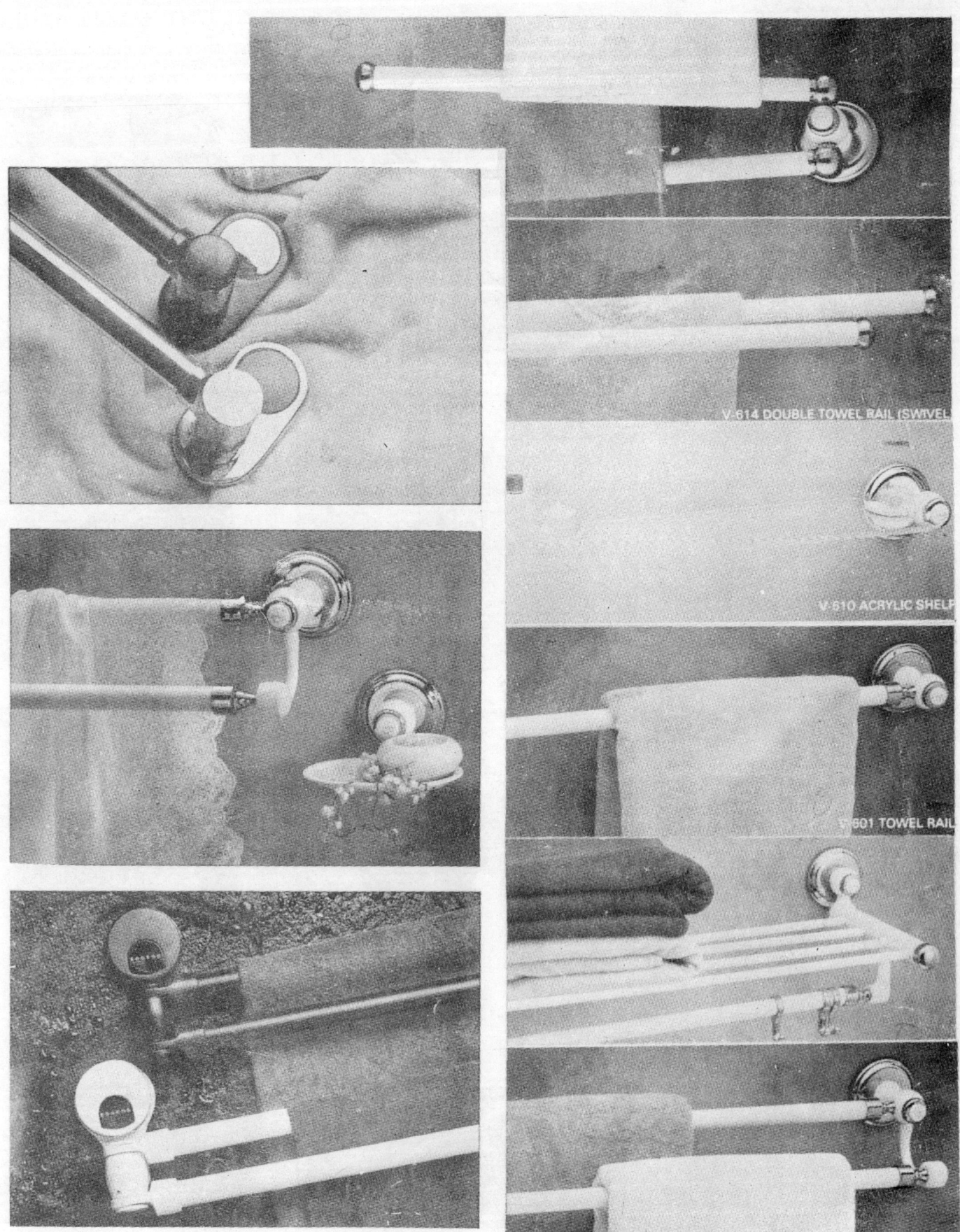

Fig. 6.8 An array of bathroom accessories.

washbasin. Mirrors are space makers that are able to give the illusion of added dimension, decorators and reflectors of light, objects of light and beauty when properly used. Many prefabricated bathroom cabinets have mirror doors, some with in-built lights, but they are not the most convenient solution, as you may want to reach the contents of the cupboard at the same time as someone else is using the mirror.

Mirrors create illusion of space, if properly used these slabs of mirror make you feel that you could walk straight through them into an unexplored dimension. To do this successfully, the mirror has to be placed where, structurally, if there were no wall, you could walk on. It is good putting one beside a window which looks out over the garden. The mirror will reflect the room, and eye will see the garden and the room in the same dimension. But placed on an inner wall where you could expect more rooms beyond, a mirror will have quite a different effect. Mirrors can be invaluable in helping to lighten and brighten dark corners in a bathroom. Two mirrors placed opposite each other butting up to the window wall, e.g. will give a room the illusion of greater width and will also make it lighter by bouncing the light further back into it. This is a very good way of brightening up a long, narrow bathroom or a dark confined area such as a separate toilet.

Glass mirrors, being heavy and difficult to work with, are being replaced by silvered acrylic sheets. The mirrored reflection in silvered acrylic is undistorted and is half the weight of glass and is practically unbreakable. It can be cut with ordinary saw and fixed with screws or with a special sticky tape. It is available in various colours and is immune to condensation. There is even a specially toughened grade which can withstand the rough treatment.

Another alternative is a plastic film with a reflective metal coating. The film is very thin, and can be mounted on a rigid surface to make a mirror, which makes it more versatile both decoratively and functionally.

Terylene treated with a special reflecting product is also available. It is stretched on an aluminium frame over a resilient backing board. The result is a gleaming, perfect mirror surface, that is unbreakable. It is easily punctured or cut, therefore, its use should be limited to the situations where this is not likely to occur. Thus, it is ideal for ceilings, its virtues for that purpose being its extreme lightness and the fact that it is immune to condensation. It is easily fixed by means of impact fabric pads. It is impervious to heat from lights, and apertures to take ceiling lights can be cut into specials. Mirrors are extensively used for decoration and sparkle rather than for space making in the bathroom. Available, for this purpose, are coloured, engraved or etched glass and framed mirrors in numerous sizes, designs and shades to give a decorative effect to the bathroom.

CHAPTER

7

BATHROOM STANDARDS

WASHING

For the efficient functioning of the bathrooms, it is essential that the spaces are designed according to the various human activities being performed in the bathrooms, like washing hands, using water closets, etc. Before designing, it is essential to know the basic standards which affect the size and layout of the bathroom. Following are the minimum standards for the space occupied during various human activities.

SPACE REQUIREMENTS

To be realistic about the eventual components of a bathroom, there must be a compromise between the appliances, fittings, etc. which are desirable and those that are possible in the available space. To operate efficiently, each component in a bathroom must be allowed a minimum set of space, and it is far better to sacrifice an extra piece of equipment than to include it at the expense of the entire bathroom. The standard appliances require a certain minimum for installation and associated activity. Although individual plans may subject these requirements to some variations, and designer or architect may be able to accommodate special requirements more compactly, these dimensions and an excellent guide at preplanning stage. These dimensions have been arrived at, keeping in mind the average size of the various sanitary appliances and basic standards for various bathing, washing and human activities (Figs 7.1A to D).

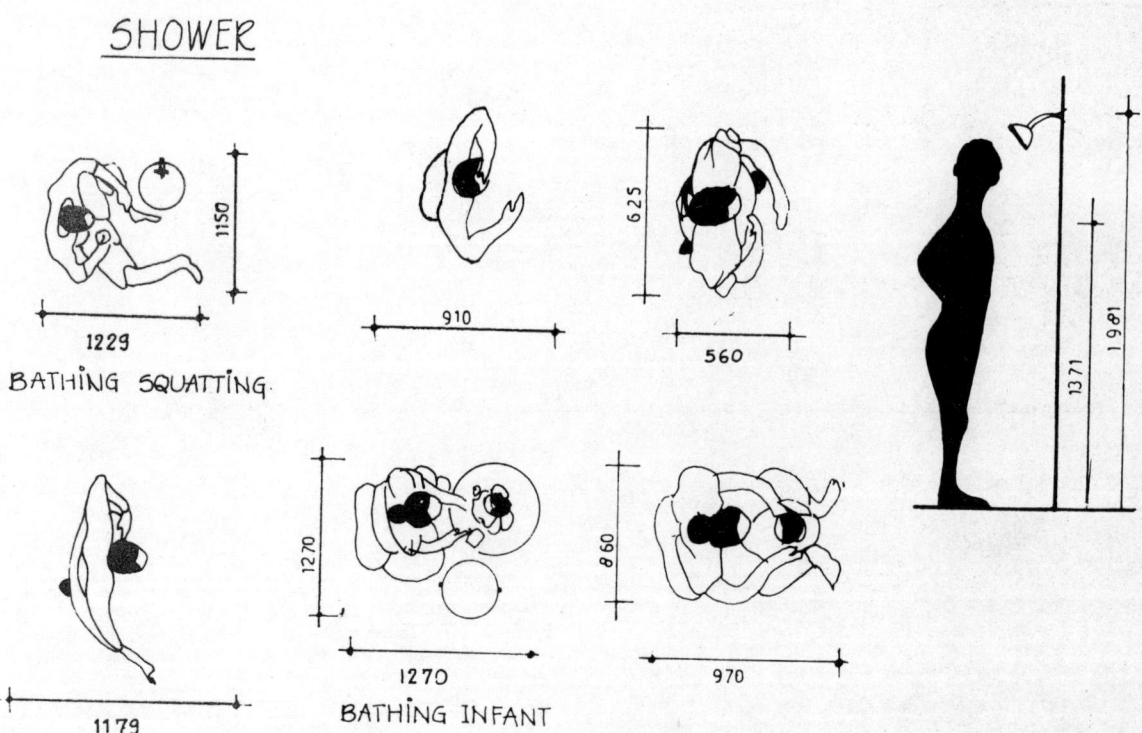

Fig. 7.1A *Details of space requirements for bathing activity.*

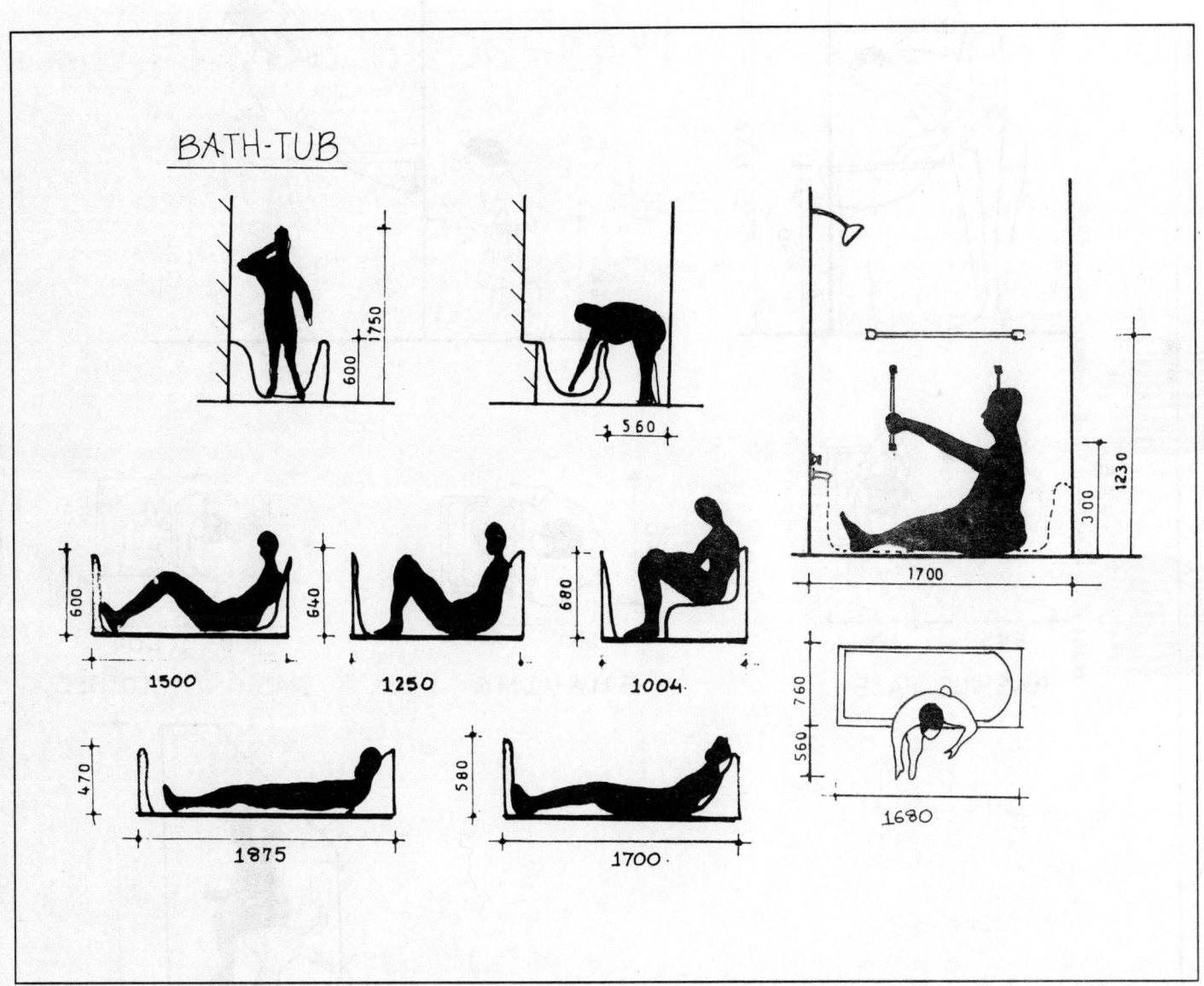

Fig. 7.1B *Details of space requirements for bathing activity.*

WASH BASIN

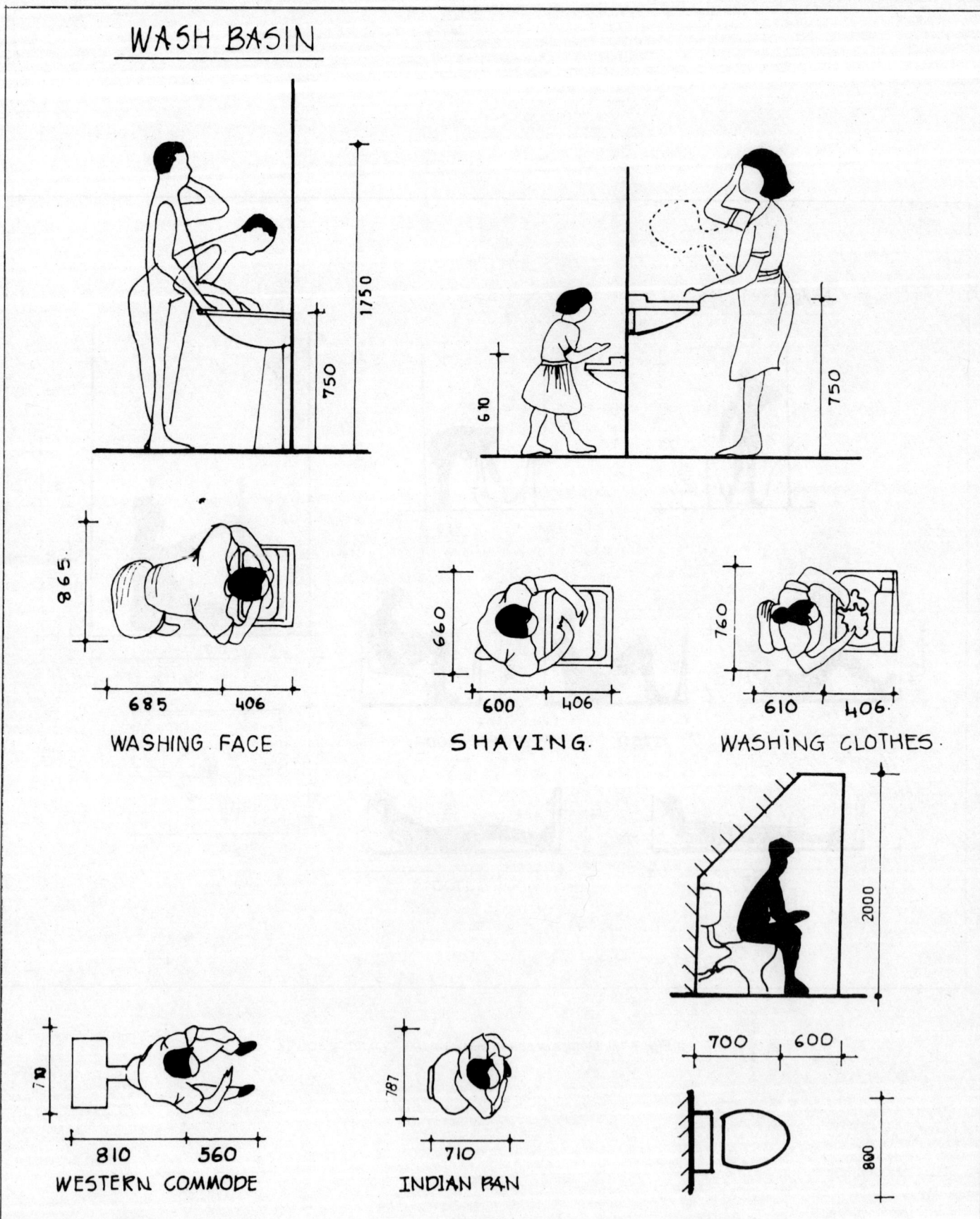

WASHING FACE

SHAVING

WASHING CLOTHES

WESTERN COMMODE

INDIAN PAN

Fig. 7.1C *Details of space requirements for activities in bathrooms for use of washbasin and water closets.*

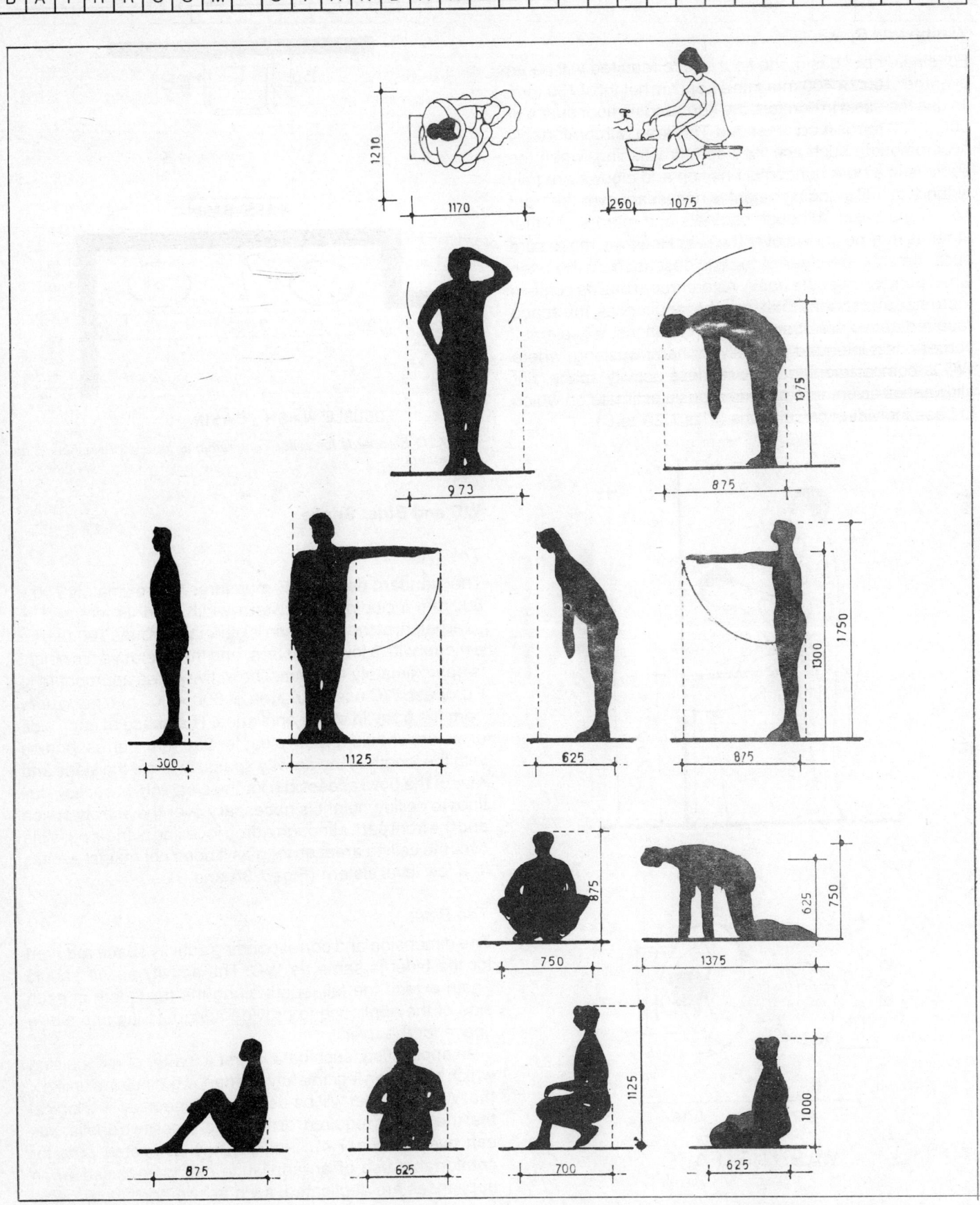

Fig. 7.1D *Details of space requirements as per human activities.*

Washbasin Space

For single bowl basin, the floor space required will be an average of 600 × 400 mm with a front rim height of 750 mm. To use the basin in comfort, the surrounding floor space of 900 × 700 mm is recommended. This area will comfortably accommodate such activities as hair washing, when the spine is in a near horizontal position and elbows are fully extended. Full standing height is necessary over the front half of the basin, although cabinets and mirrors or other fitments may be placed over the rear. However, make sure such fitments are clear of the arc described by the user, when bending over the basin. As bathroom basins come in more size and shapes than the other appliances, the space required above are open to some variations, e.g. a small corner basin intended primarily for hand washing, where WC is compartmented, requires less activity space, but allowances given here provide a basic estimate on which to base individual calculations (Figs 7.2A to C).

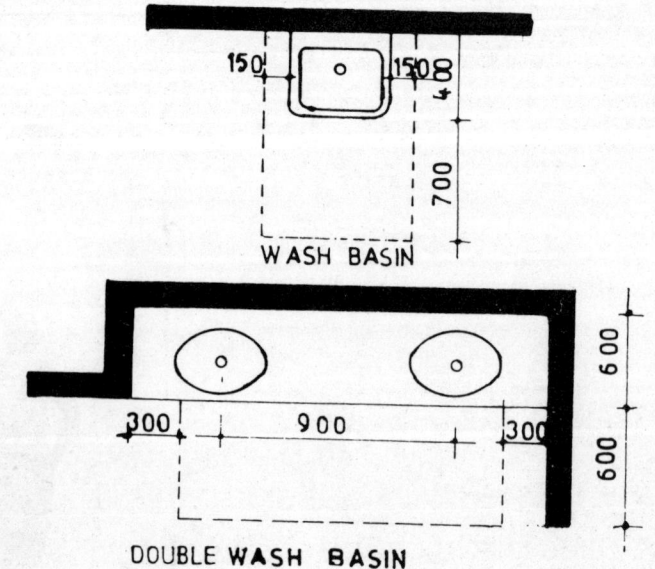

Fig. 7.2C Standards for space requirements for use of washbasin as in counters.

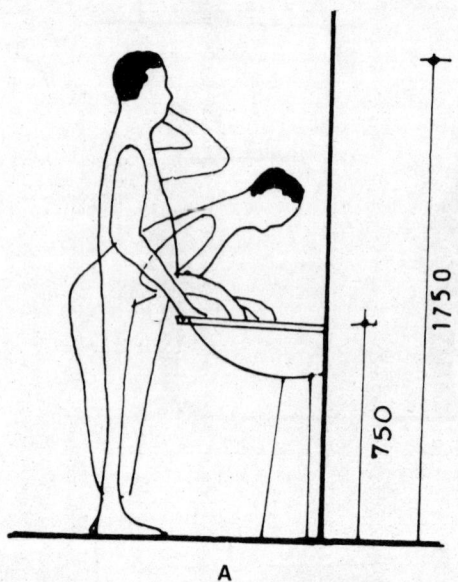

A

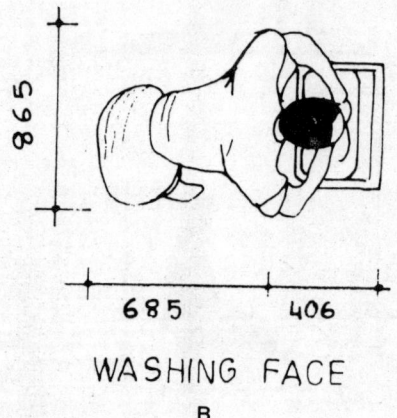

WASHING FACE

B

Figs 7.2A and B Standards for space requirement for use of washbasin.

WC and Bidet Space

The WC

The standard design WC measures approximately 700 × 800 mm including the cistern width, and its length. The total height of the appliance is generally around 700 mm for a model with a low set cistern, and the operative rim height is approximately 400 mm. The activity area appropriate to European WC use is an area of 600 × 800 mm measured from the front lip of the appliance. The space at each side of cistern should remain clear for 800 mm, corresponding with the length of the activity space, to leave the sides and rear of the bowl accessible for the cleaning. In section, full floor to ceiling height is necessary over the activity space and the front part, although a drop in ceiling plane is possible over the ceiling area, as long as it does not restrict access to a low level cistern (Figs 7.3A and B).

The Bidet

The dimension and corresponding activity space required for the bidet is same as WC. The activity space should again extend the full length along the full length of each side of the appliance to provide adequate leg and elbow space for the user.

In application, such data is first a matter of recognising which areas can legitimately overlap, e.g. since it is unlikely that WC and bidet will be used simultaneously, it is logical that the spaces required for the use of these two appliances can overlap (Fig. 7.4). Similarly, the estimated area for comfortable use of an appliance can be reduced when appliances are duplicated, such as two appliances occur on two different horizontal planes (e.g. a basin activity space can overlap that if WC or bidet by 200 mm without obstructing the use of either appliance.

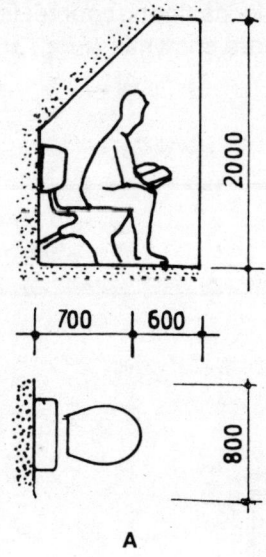

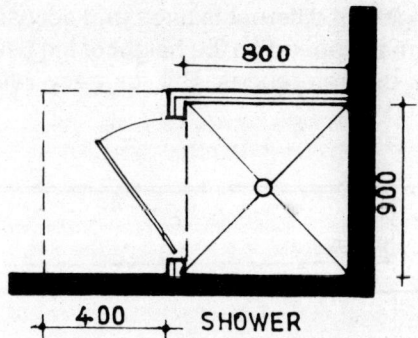

Fig. 7.5 *Standards for space requirements for use of shower.*

partially within the shower). If the shower is enclosed on three sides, the activity area need to be increased to 700 × 900 mm.

Bathtub Space

The average dimensions of a bathtub are 1700 × 700 mm with a rim height 500 and 600 mm (Figs 7.6A and B). A clear floor space of 1100 × 560 mm is required in front of the tub for—an adult to get in and out of the bath in comfort, and 1700 × 700 mm for drying and for such secondary activities as an adult bathing a child from beside the tub (Fig. 7.7). The longer dimension of the space is adjacent the tub.

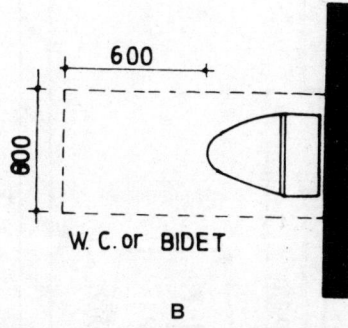

Figs 7.3A and B *Standards for space requirements for use of European water closets.*

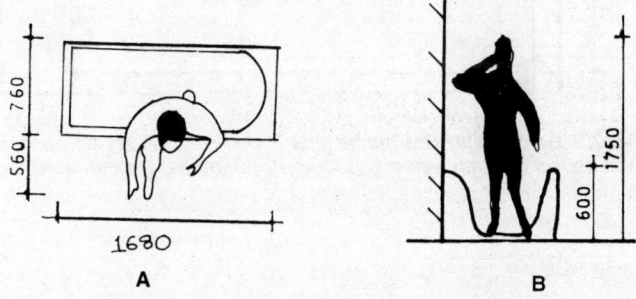

Figs 7.6A and B *Standards for space requirements for use of bath tubs.*

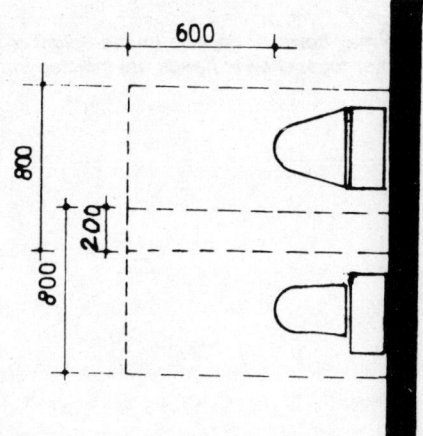

Fig. 7.4 *Standards for space requirements for use of bidets.*

Shower Space

The space required by the shower appliance is 800 × 900 mm (Fig. 7.5). If the shower is enclosed on two sides, the activity area to be left in front of the shower should be at least 400 × 900 mm (or the width of the tray) for suitable access and room for drying (the space can be

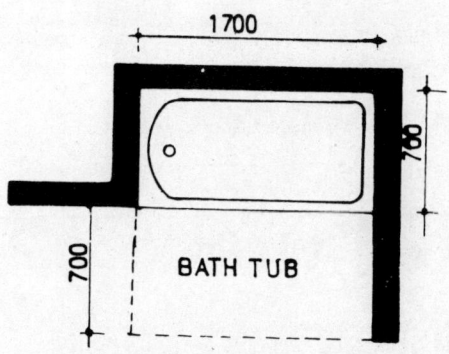

Fig. 7.7 *Standards in regard to circulation area around bathtub.*

The heights of different fixtures and accessories in the bathroom may depend on the height of the user as well as his or her or preferences. But for easy reference, the standard heights for cabinets, counter tops, shower heads and accessories are shown in (Fig. 7.8).

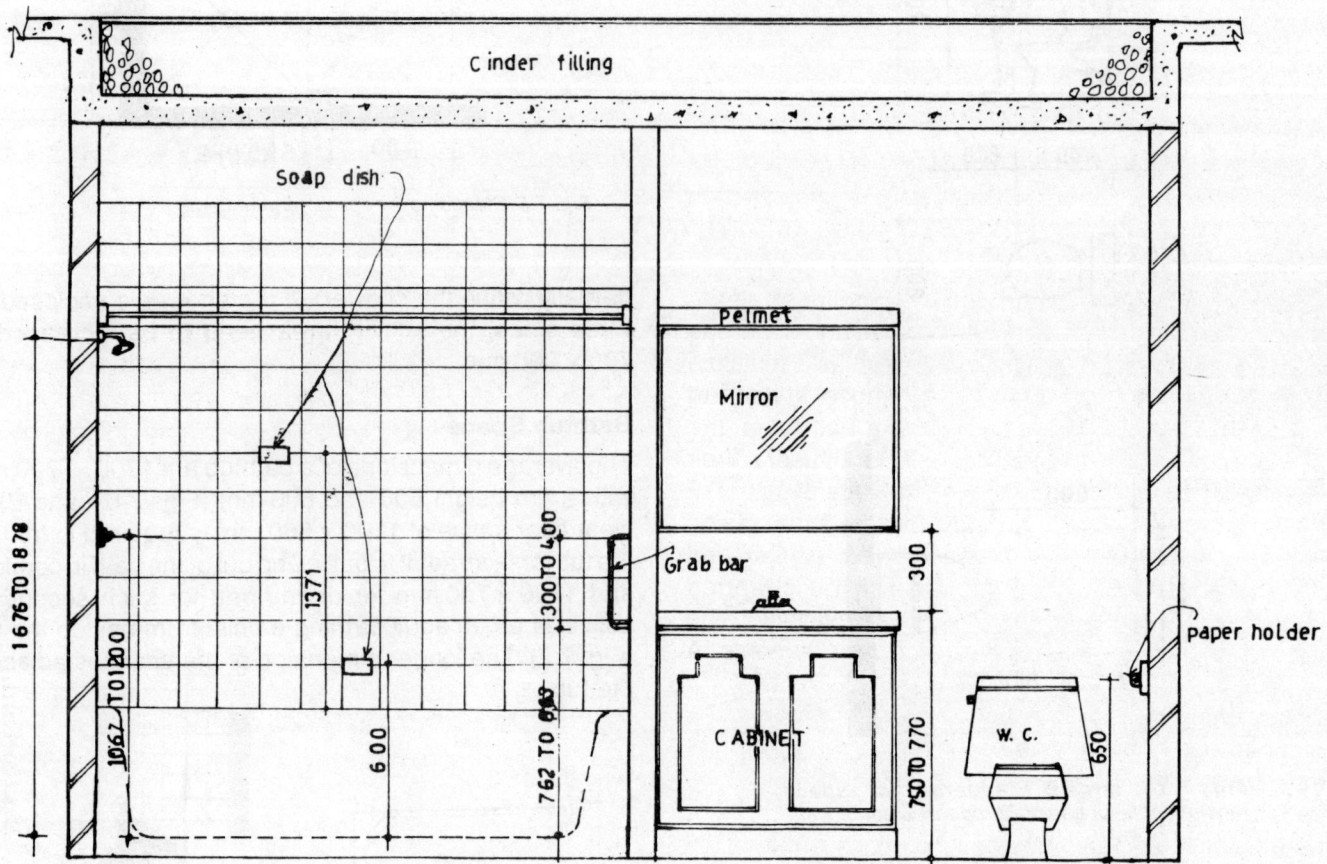

Fig. 7.8 *Section showing the heights of different fixtures and accessories in the bathroom may however depend on the height of the user as well as his/her preference. But for easy reference, the standard heights for cabinets, counter tops, shower heads are indicted in the sketch.*

CHAPTER

8

BATHROOM LAYOUTS

The definition of an ideal bathroom is, a functional, clean, maintainable and tidy appearance, with wet and dry areas properly segregated. Designing a bathroom is a multifold process—planning the space, defining a style and choosing components. While designing a bathroom, the most important thing is to know the number of users, their ages and their requirements. Whether a place of luxury or utility, bathroom involves arrangement of three basic elements, i.e. washbasins, water closets (WCs) and bathing space. Their arrangement can be changed according to the requirement and space available. The first towards planning is to assess the requirements. A good, efficient design cannot be evolved unless we know the things we require in the bathrooms. Bathrooms can be planned according to the function, i.e. family bathroom, guest bathroom or children bathroom. Some considerations will be common to all categories, others dependent on the activity you envisage in the bathroom in question, e.g. bath, basin and WC are common to all the bathrooms. Ventilation, soundproofing and lighting must receive equal attention. Beyond that the requirements for each room may vary considerably, in practical as well as decorative terms. Although adults or near adult children use the bathroom more efficiently than younger family members, they tend to take more time about it. Such procedures as shaving and general beauty routines are the province of a bathroom. When several routines must be accommodated, they become accountable factors in planning for efficiency. Secondary items such as a separate shower cubicle, bidet, open storage space, easy chairs, exercise equipments, etc. can also be included. Decorative effects, too, can be of a more lavish nature, since they are not so prone to wear and tear.

Bathrooms are basically arrangement of three elements—WC, washbasin and shower. But still, with the advancement of technology and users attitude of making it a luxury and place of relaxation, it has become necessary to design a bathroom in such a way that functionally and aesthetically good unit is obtained. The basic layout depends on the user's requirement, e.g. few people want a separate WC and bath, sometimes combined unit is required. The following are the few basic layouts.

SEPARATE WC

Separate WC is a two fixture room, also known as guest bath powder room. This room contains a WC and a wash-basin and perhaps some limited storage space, where hand towels and some toiletries can be stored for the use of the guest. Fixtures can be placed side by side or on opposite walls, depending on the shape of the room and the space available (Figs 8.1A to D).

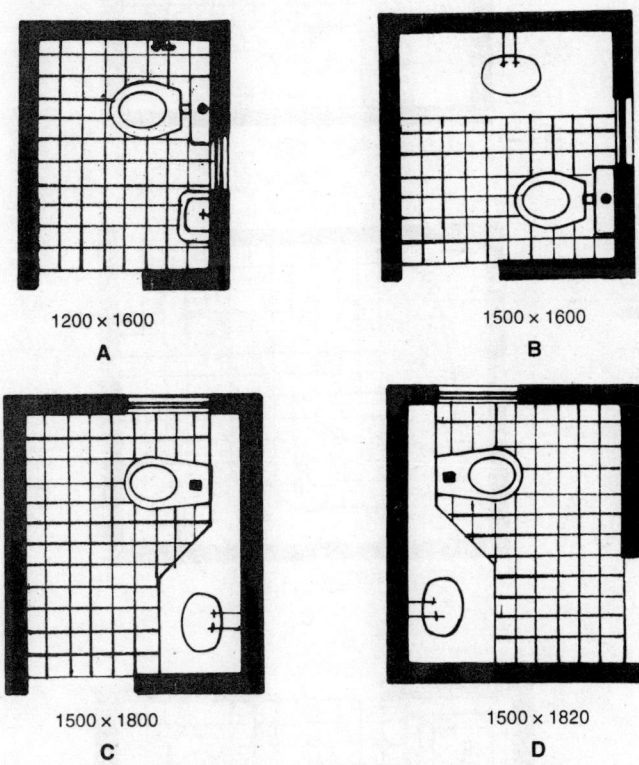

1200 × 1600
A

1500 × 1600
B

1500 × 1800
C

1500 × 1820
D

Figs 8.1A, B, C and D Guest bath/powder room.

UTILITY BATH

The utility bath is a functional room and only large enough to accommodate three basic fixtures—a WC, washbasin and a bathtub or shower. Material of fixtures and finishes should be durable and easy to maintain. Little storage space for toiletries and cleaning material is also provided either in the form of shelves, niches or small cabinets (Figs 8.2A to D).

BATH WITH DRESSING AREA

Some bathrooms are provided with an adjoining dressing area for the quick access to clothes after bathing. The dressing area includes closet for clothes and dresser with mirror and counter for cosmetics. In addition to closet and dresser, the dressing area should be big enough so that there is sufficient space for dressing up. This area should be well lit for make-up application (Figs 8.3A to C).

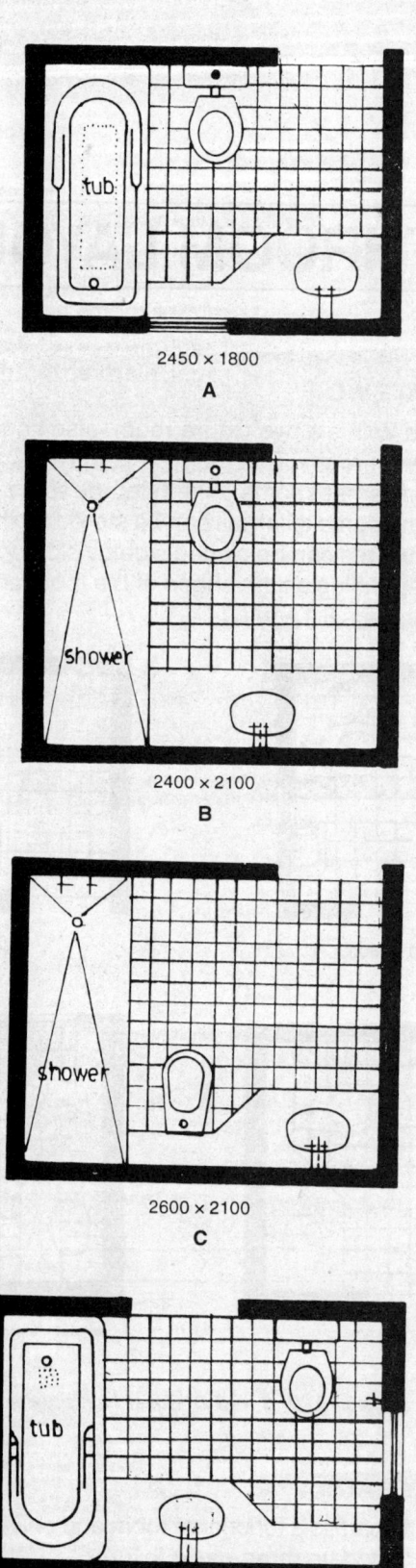

Figs 8.2A, B, C and D Different designs of utility bathrooms.

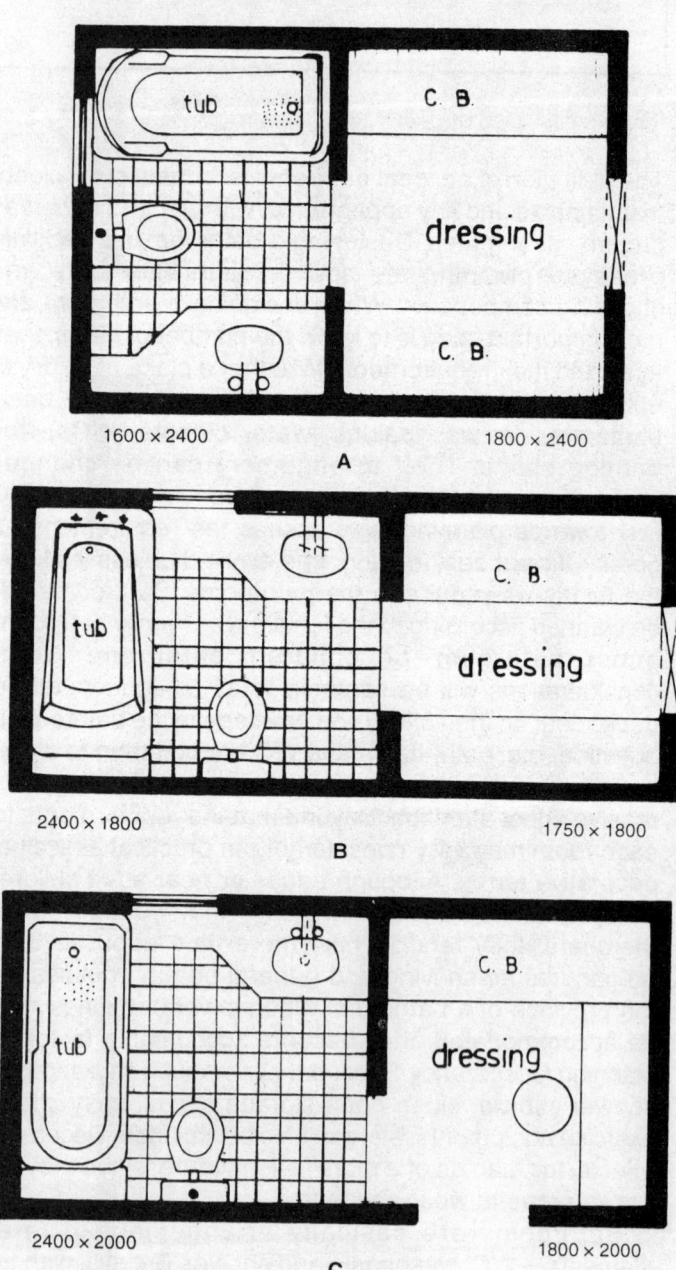

Figs 8.3A, B, C Different designs of bath with dressing area.

LUXURY BATH

The luxury bath is more than just a place to grab a quick shower and run a comb through your hair. In addition to the three basic fixtures—WC, washbasin and bathtub, bidet and separate shower enclosure are also sometimes included (Figs 8.4A to C). These bathrooms have a sufficient storage space. The aesthetic part of the bathroom is given due consideration with matching fixtures and finishes. The mood and accent is created with the use of lights. This bathroom is designed for relaxation at the beginning and end of the day.

SUPER LUXURY BATH

Super luxury baths are not merely a utilitarian space, this bath reflects the personality and interest of its owners. It includes dressing and grooming areas, WC and bathing facilities, and other amenities—Jacuzzi baths, bidets, steam/sauna bath, and exercise rooms are often included in contemporary design (Fig. 8.5).

Here are some more one can plan into the bath (Figs 8.6 to 8.8).

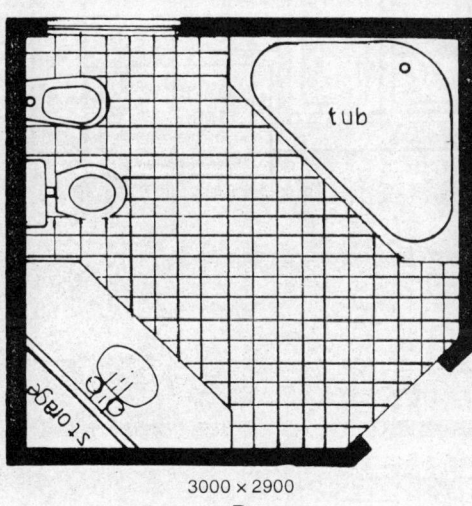

2700 × 2900

A

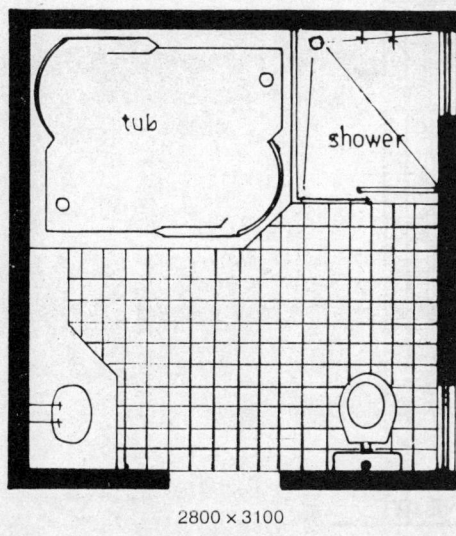

3000 × 2900

B

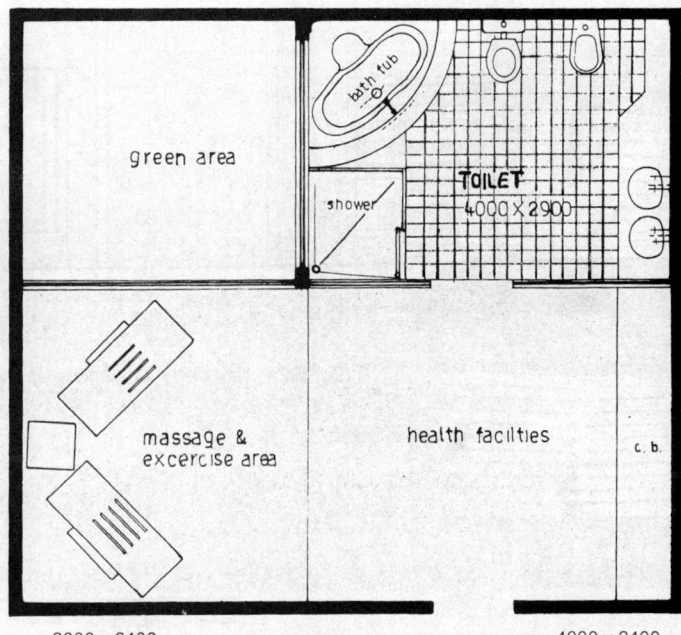

3000 × 3400   4000 × 3400

Fig. 8.5 Design of super luxury bath.

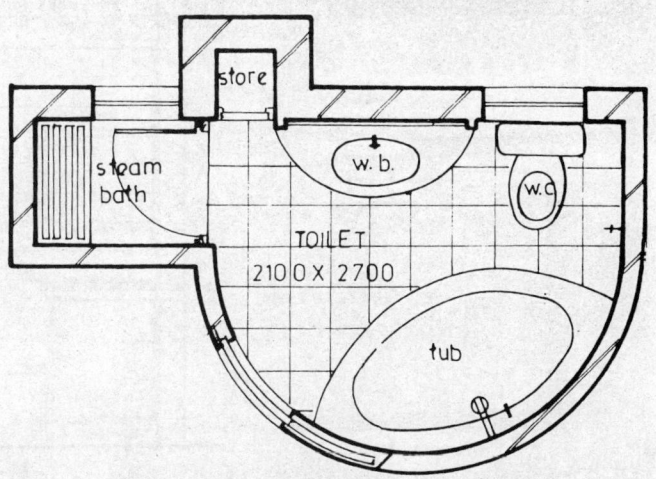

2800 × 3100

C

Figs 8.4A, B, C Different designs of luxury bathrooms.

Fig. 8.6 Plan layout of a luxury bathroom.

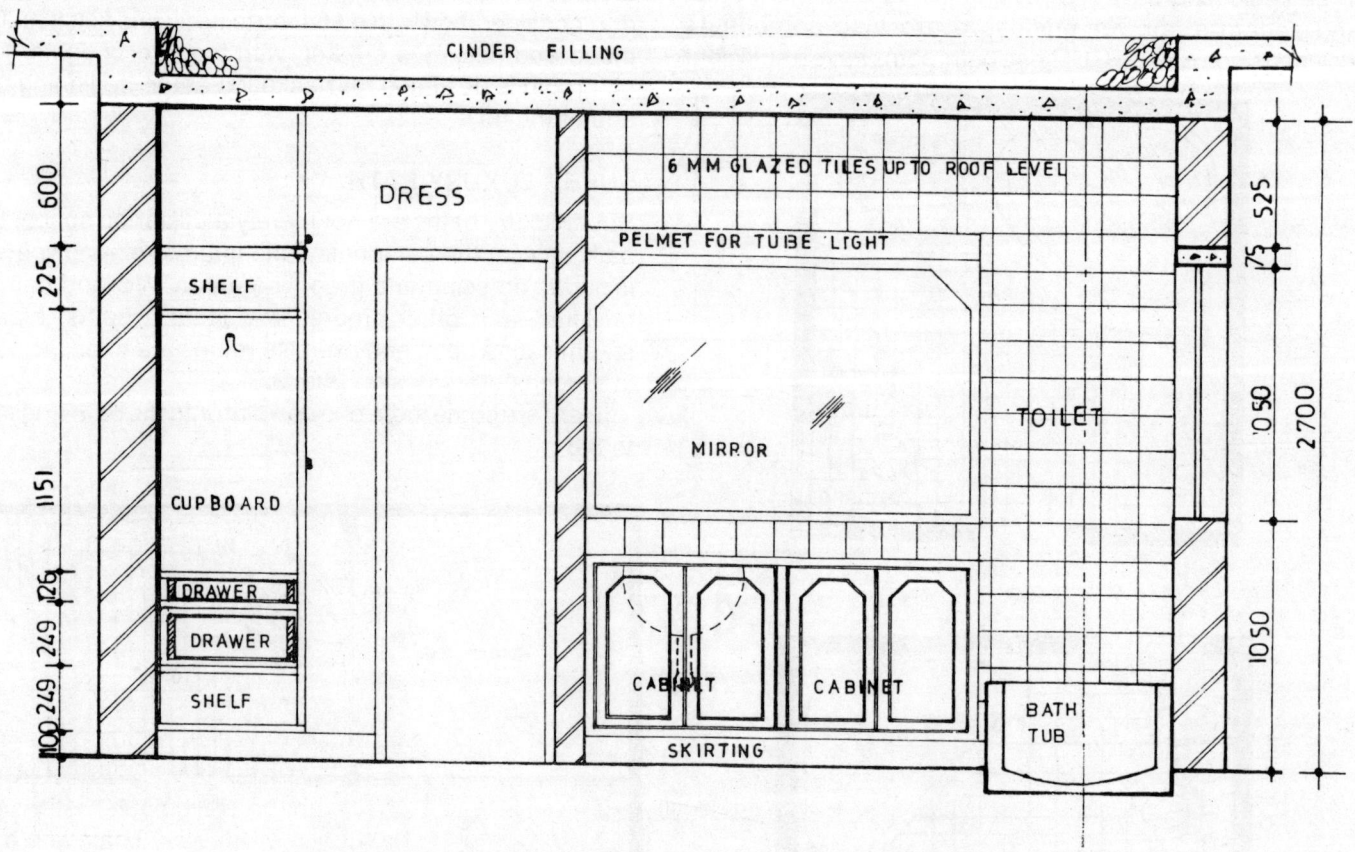

Fig. 8.7A *Working plans and section of toilet cum dress showing various details.*

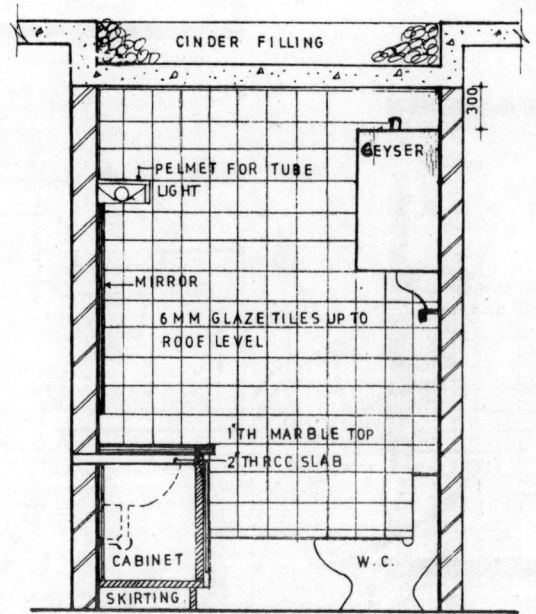

Fig. 8.7B *Working plans and section of toilet cum dress showing various details.*

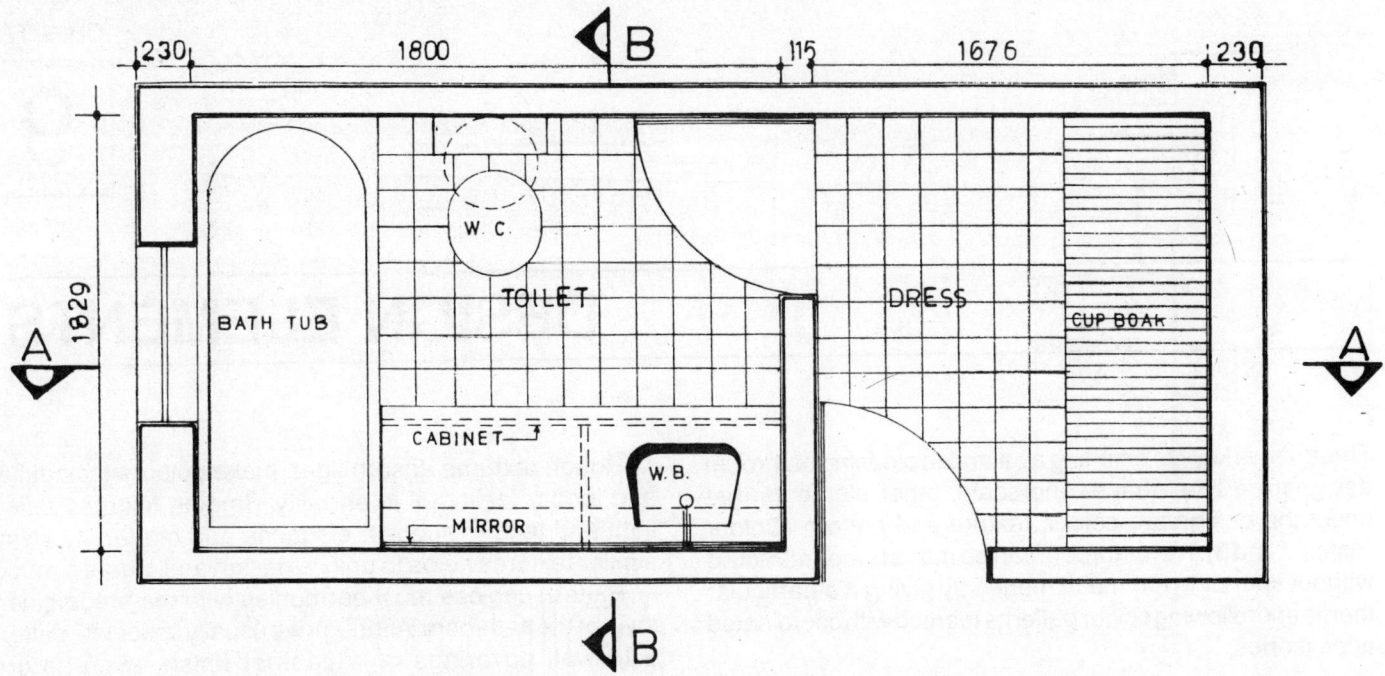

Fig. 8.7C *Working plans and section of toilet cum dress showing various details.*

Fig. 8.8 *View of the luxury bathroom.*

Make sure that adequate ventilation is provided to prevent water damaging delicate objects and equipments.

Reading nook This might be the quietest refuge in the house. a cushioned bench or an easy chair with a good light source (may be a skylight) is all one needs for a few minutes of solitude. A small bookcase or magazine rack will be useful.

Art gallery One can show works of art, build craft pieces, such as hand-made tiles or a stained glass window, right into the design.

Greenhouse Because of the high moisture level, plants often thrive in a bathroom. It is an ideal place to bring a touch of nature into the house. If somebody does not want the plants inside the bathroom, the bathrooms can overlook a landscaped open space with skylight on top. The bathroom can have a glass wall to have a clear view.

CHAPTER

9

DESIGN ELEMENTS

Three visual keys to planning a balanced pleasing bathroom designs are line, shape, and scale, other elements that affect the design are colour, texture and pattern (Colour Plates 7 and 8). The bathroom can be made to look attractive without spending too much money by giving it a particular theme and following colour patterns teamed with coordinated accessories.

LINE

Vertical lines give a sense of height, horizontal impart width, diagonals suggest movement, and curved and angular lines contribute a feeling of grace and dynamism. Continuity of lines unifies a design lines created by shower or tub unit, cabinets, counter, windows, doors and mirrors fit together. It is not necessary that everything should align but if a number of elements, especially the highest features in the room create a continuous line—the effect is quite pleasing.

SHAPE

Continuity and compatibility in shape also contribute to a unified design. The shapes formed by doorways, windows, countertops, fixtures and other elements like patterns in flooring, wall covering, shower curtains, and other accessories should be compatible.

It is not advisable to repeat the same shape throughout, they may be monotonous and boring. They should be only complimentary and compatible.

SCALE

The scale of the bathroom elements have to be in proportion to the room to make the design harmonious and pleasing. A small bath seems even smaller if equipped with large fixtures and a large counter. The same bath can look larger if filled with space-saving fixtures, and open shelves. When wall cabinets or linen shelves extend to the ceiling, they often make a room seem top heavy—and therefore smaller.

TEXTURE AND PATTERN

Textures and patterns work like colour in defining a rooms style and sense of space. The bathroom's surface materials may include many different textures—from a glossy countertop to wooden cabinets to a quarry tile floor.

Rough textures absorb light, make colours look duller, and lend a feeling of informality. Smooth textures reflect light and tend to suggest elegance and modernity. Using similar textures helps to unify a design and create a mood.

Pattern choices must harmonise with the predominant style of the bathroom. Although we usually associate pattern with wall coverings or a cabinet finish, even natural substances such as wood, brick, and stone create patterns.

While variety in texture and pattern adds interest, too much variety can be overstimulating. It is best to let a strong feature or dominant pattern be the focus of the design, and the other surfaces can be selected to complement it.

COLOUR

Before selecting any colour scheme for the bathroom, it is important to know the basics of colour theory and the colour wheel (Fig. 9.1A and B, *see* Colour Plate 9).

Colour theory is not just the painting of the colour wheel but includes understanding of value and how to use it and control it, using and controlling intensity, colour schemes, line of designs, forms, background treatments, light and shadow, centre of interest, positioning colour, etc.

Basic Colours

- The basic hues as shown in the colour wheel (Fig. 9.1A, *see* Colour Plate 9) are vivid or pure colour.
- Light colours are the mixtures of the basic hues with white, which lessens the intensity of the colours. Also known as tint of the colour.
- Dull colours are mixtures of the basic hues with grey, which tend to muddy the colours. Also known as tone of the colour.
- Dark colours are mixtures of the basic hues with black. Also known as shade of the colour.
- Achromatic colours are literally "colours without colour"—in other words, black and shades of grey.

Every colour in the spectrum has different characteristics. *Red* is passionate, the colour of hearts and flames, it attracts the attention, and actually speeds up the body metabolism. Red is popular among young, and pink in particular is associated with romance. Deep red looks aristocratic.

Yellow is lively and happy, the colour of sunshine. Because it is so relentlessly cheerful, we tend to be tire of it quickly. Pale yellow makes room breezy and spring like.

Green is the colour of trees and grass. Bright green reminds us of spring and fertility, but it is also the colour of mildew, and jealousy.

Blue is the colour of the sky and the sea. Like green, it has the colouring effect, but it is also quite powerful—the strongest of the familiar colours after red. Light blue looks young and sporty, but the royal blue and navy blue have dignified wealthy air.

Purple is the sophisticated colour, long associated with royalty. We do not often see it in nature, so we think of it as an artificial colour, and find it a bit hard to take.

Brown is rich and fertile, like soil, and it is also sad and wistful, like the leaves in autumn.

Black is the colour of night and death, and is often linked with evil. Its unorthodox appearance has made it popular with artists and designers, but it is also associated with wealth and elegance.

Warm Colours

The hues from red to yellow are called the warm colours (Colour Plate 9). Warm colours are bright, splashy and aggressive. More than any other colours, they attract the eye and excite motions.

Cool Colours

The hues from green to violet, including blue and all the shades of grey are known as cool colours—perhaps because they remind us of snow and ice (Colour Plate 9). Cool colours have exactly the opposite effect as warm colours, i.e. they slow down the body's metabolism and are even used in hospitals to calm maniac patients.

Light Colours

Light shades of any colour look soft and ethereal. The hue is relatively unimportant. Light colours overwhelmingly preferred in interiors designs—dark colours tend to make a room look gloomy, but it lacks the eye-grabbing quality.

Dark Colours

Black and other dark shades feel heavy. The dark colours are used mostly for type and accent and are usually paired with lighter, more conventional colours.

Vivid Colours

Vivid colours have powerful personalities when you combine two or more vivid colours, the result is chaos. Use at least one vivid colour as an accent in colour scheme. Vivid colours must be used sparingly. In general, use a vivid colour as the accent colour and a light or dull colour for the background.

If you need another accent colour, use a darker shade of the vivid colour. Vivid colours like red and blue are sometimes used for background. But there is almost no way to mix two vivid colours without mediating them with as neutral or achromatic shade.

Choosing Colours for Bathroom

Before you even think about which colours to choose, you should be familiar with the three characteristics of colour (hue, lightness and saturation), the uses and emotions associated with each colour, and the six broad categories of colour mentioned earlier (warm, cool, light, dark, vivid and hue).

Any colour scheme should reflect the personality of the user. Before choosing any colour scheme, ask yourself the following questions:

- What kind of effect do I want?
- What colours will best convey this effect?

Choose the background colour first for the walls and the ceilings, then pick a darker or more vivid shade for the sanitary appliances, bathroom accessories, mouldings, for mirrors, etc. When choosing a background colour, light colours work better than dark colours, warm colours are better than cool colours.

When choosing a colour scheme, pick the shades first, not the hues. If you decide to vary shades, it is important to find some common ground between colours, and this usually means hues that are similar or otherwise compatible. You can even use contrasting or complimentary hues as long as there is not too much contrast in shade, but be careful when using colour that clashes, like red and green or red and blue.

Another very good way to increase colour harmony is to limit the number of colours in colour scheme. Two or three colours are usually enough, five is too many—nothing looks worse than too many colours, particularly when they lack common elements.

By choosing the light colours and thereby reflecting the optimum degree of light, a small room can be made to look larger and *vice-versa*.

Light colours reflect light making walls recede, thus, small bath treated with light colours appears more spacious. Dark colours absorb light and can visually lower a ceiling or shorten a narrow room. When considering colours for a small bathroom, too much contrast has the same effect as dark colours, it reduces the sense of spaciousness. Contrasting colours work well for adding accents or drawing attention to interesting structural elements. But if a problem feature in the bathroom is to be concealed, it is best to use one colour throughout the area.

Warm or cool colours balance the quality of the light, depending on the orientation of the bathroom while oranges, yellows, and colours with red tone impart a feeling of warmth, they also contract space. Blues, greens, and colours with a blue tone make an area seem cooler and larger. A light monochromatic colour scheme (using different shades of one colour) is restful and serene. Contrasting colours add vibrancy and excitement to design. But a colour scheme with contrasting colours can be overpowering unless the tones of the colours are varied.

LIGHTING

Lighting is a part of the initial designs. A beautifully designed place can also be unpleasant due to inadequate lighting. Since, bathroom is a multiple use room, different lighting levels are required for different tasks, which vary from very soft ambient light to strong directional task lighting.

The lighting requirements of your bathroom will differ depending on its size—whether it is small or large, and also on the need of people who are using it. A pleasing level can be achieved by the natural light, from windows and skylights, supplemented by artificial lighting. Lighting can be divided into three categories: task, ambient, and accent. Task light illuminates a particular area where visual activity takes place such as shaving, applying make-up, etc. Ambient lighting fills in a bathtub. Accent lighting is decorative, and is used to highlight architectural feature.

Natural Light

A single window in the middle of a wall create a glaring contrast with the other feature of the bathroom. Windows can be placed on adjacent walls or a skylight can increase the amount of natural light.

Glass can be used in different forms (panels and glass blocks), etc. and finishes (clear and translucent) to bring in light and view still ensuring privacy. If plain glass (transparent) is to be used, sill level of the window should be above chin height of the tallest occupant. But before that, the heights and locations of the neighbours' window should also be checked.

Artificial Lighting

Task, ambient and accent can be created by artificial lighting. Lighting in a bathroom should be small and discrete ceiling lights are better as they do not cast shadows. Keeping task and ambience in mind, the light should be bright over the washbasin/counter medium in the bath area and subdued over the WC. The light on the washbasin should throw light on the person's face rather than on the mirror surface.

A great variety of lights and fixtures are available in the market today.

Incandescent Light

Incandescent light is the most commonly used fixture—traditional light bulbs also known as tungsten light, low voltage incandescent lights are good for accent lighting.

Fluorescent Light

Fluorescent light or what we call tubelights, are energy efficient and good for general lighting in the bathroom. They give shadow-free light. Some designers do not approve use of this light above mirrors as it produces a harsh tone giving a greyish tinge to the complexion.

Quartz Halogen

Quartz halogen good for task lighting, but its initial cost is very high and it is emits a lot of heat. They produce a bright white beam and are used in places where you need concentrated light. Low-voltage lights are very good for accenting particular features in the bathroom.

Task/Activity Lighting

Shaving, putting on make-up are the activities that require very good light, so, the area around the mirror and the washbasin has to be well lit. Light from both sides of the face, while looking into the mirror is the best. The overhead light on the mirror is the best. The overhead light on the mirror distorts and casts shadow on the face. A good arrangement is to place bulbs around the edge of a mirror. The warm tones of these bulbs give accurate skin colour. But care should be taken that the bulbs are properly sealed against water. A tungsten strip behind a casing is an effective form of light and should be long enough to cast light on the sides of the face as well as front. Ceiling lights may produce shadows on the face which can be avoided by placing fixtures close to the wall on which the mirror is mounted. For those who are in the habit of reading in the bathroom, a strong reading light should be provided above the WC and the bathtub.

Ambient/General Lighting

In a small bathroom, a single tungsten bulb or fluorescent light may be enough to light up the whole room. The bulb should be away from reach of water.

Mood/Accent Lighting

Lighting is a mood setter and can provide the perfect aid for unwinding after a stressful day. The effect could be bright and fresh for the day, subtle for long luxurious baths and sparkling with a touch of glamour for a sophisticated look. Display art on your walls and highlight it with special lighting. Spotlights highlighting architectural features, art objects or plants can transform your bathroom. Neon lights on the ceiling can give a special effect to the bathroom. By accenting special features in the bathroom, attention is diverted from the toilet and fittings, and the bathroom attains a character of its own.

Bathroom lighting should combine efficiency with safety. Great care is must to ensure that the light switches are not accessible from the shower area. It is advisable to place all plug points as far away from water outlets as possible to avoid accidents. All light fittings should ideally be sealed. There are fixtures available in the market, specially made for the bathroom with safety in mind.

VENTILATION

Condensation and the humid steam are not particularly comfortable as these tend to render mirrors temporarily useless as well as adding to the general cleaning problem. If one of the walls of the bathroom is an external wall, the provision of proper ventilators can solve this problem. But in case of internal bathrooms, a grating and a duct to the open air or mechanical ventilation is the solution.

A mechanical ventilator is a priority item for all bathrooms. Open windows are rarely welcomed in winters, and they are useless as ventilators when closed. For a small bathroom, a small electrically operated extractor fan, ducted and sound insulated in an internal bathroom will suffice in a minimum area and is more economical with a manual switch to be operated only when necessary (Fig. 9.2). In large bathrooms, the larger, more sophisticated ventilators which are wired to the lights and operate automatically when the lights are switched on are more efficient. Auxiliary exhaust vents in the bath enclosures and shower compartments also help to reduce the condensation level. A fan near another exterior opening should be avoided as the air will not circulate in the room.

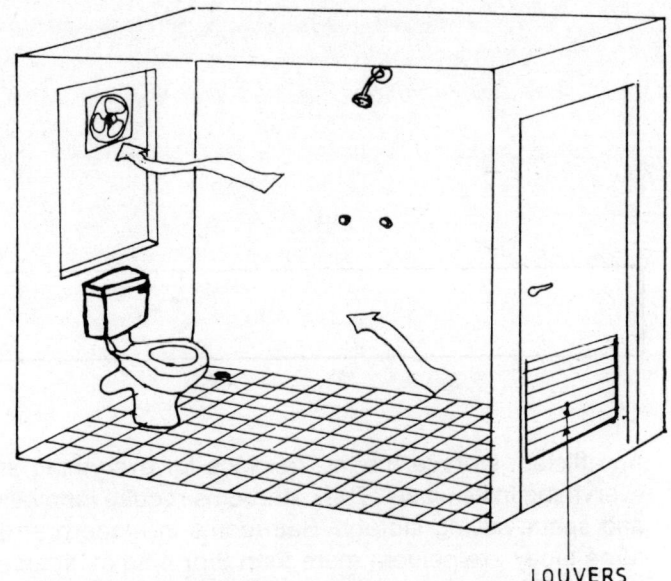

LOUVERS

Fig. 9.3 Small bathroom with or without a small window can always use an exhaust fan for ventilation and exhaust.

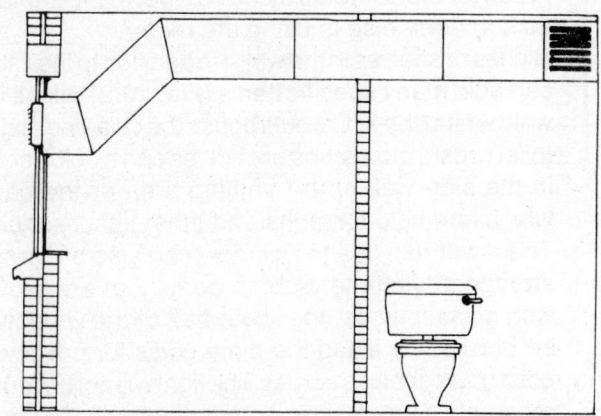

Fig. 9.2 Ventilation-extractor fan and ducting to change air in a windowless bathroom.

Small bathrooms, with little or no window ventilation, can always use an exhaust fan in addition to the regular ceiling or wall fan (Fig. 9.3). It is also useful to have a louvered bathroom door with correctly directioned louvres in the lower portion. This will let in some air, and may even facilitate a cross-flow, while keeping the much needed privacy.

SOUND INSULATION

One of the most important considerations in locating a bathroom is the noise transference factor between rooms. Transference noise is referred to either as impact noise transmitted through structural elements or air-borne noise. The first is not such a problem, since it can be controlled by insulating, and by keeping appliances away from partition walls. It is the air-borne noise which will mostly affect the location of a bathroom. Since, where it is likely to be a problem, the only really effective cure is an extra heavy partition. This may be structurally costly. Thus, a realistic appraisal of the noise generated in bathroom and the surrounding rooms at planning stage is important. There are several other remedies—one is to have a lobby between the bathroom and living areas. Built-in cupboards, lining the walls which adjoin either the bedroom or the corridor will do a considerable muffling job. Even the fact that walls are lined with tiles or wood panelling will help and so will a quantity of sound absorbing soft surfaces in the room, such as carpet, curtains, acoustical ceiling tiles. At the very least, make sure you have a heavy tight-fitting door through which the minimum of sound will penetrate.

The other sort of noise comes from the plumbing and flushing cistern which can be irritating if you are a light sleeper in the next room. The siphonic flush water closet functions much more quietly than the washdown type. The pipes need to be of correct size and securely fixed to the adjacent walls.

CHAPTER

10

STORAGE

An efficient storage means, "a place for everything and everything in its place". The bathrooms require innovative and space-saving storage. Bathrooms in modern structures today are seldom more than 3 or 4 sq m, scarcely enough to house the three basic fixtures—WC, washbasin, and the shower area of bathtub. Thus, the good design is to provide the sufficient space for storage and at the same time having an uncluttered and clear spaces in the bathroom.

While still in the planning stage, storage should be given a proper thought. First of all, think of all the things one would like to have in the bathroom and create the niches, shelves, drawers or cupboards for them for the efficient use of the bathroom. Adequate storage should be provided for current and reserve supplies.

Articles in current use should be located near their place of first use. An accurate list of the objects to be stored is necessary for the scientific allotment and arrangement of space and facilities. A margin of safety of some 25 per cent increased capacity should be allowed for the usual accumulation of additional belongings. Apart from the personal toiletries, extra cakes of soap, bottles of shampoos and conditioner, spare toothbrushes and toothpastes, face and toilet tissues, detergents and cleaning materials need a room in the bathroom itself. One needs place for hand and bath towels, bathrobes, nightwears, etc. and logically, the bathroom is the best place to stock them, close at hand where they might be needed in a hurry. Dirty linen and clothes are also required to be dumped in the bathroom. These days dressing areas have also shifted to the bathrooms. Thus, you need a space for make-up and cosmetics also near the mirror. That includes the cupboard for clothes as well in the bathroom. Now the good design is to fit all these things, may be hidden from the view, compromising in the clear spaces.

After the list of the items to be stored in the bathroom has been decided, decide on the approximate shelf area required (allowing a bit extra which although luxury at the start soon becomes the necessity). The items should be placed near its area of use. There are many ways by which the storage can be provided in the bathroom depending on the size of the bathroom, the convenience of the user, etc.

In small bathrooms, where space is a problem, following are few of the solutions to provide storage area.

1. A cabinet with a mirror door over a washbasin is ideal for small bathrooms (Fig. 10.1). These cabinets are available in the market with or without the in-built light. These can contain the toiletries like toothbrush, toothpaste, soap, shampoos, etc. These cabinets are not the most convenient solution as you may want to reach for the contents of the cabinet at the same time as someone else is using the mirror.

2. Niches or shelves in the walls, in addition to the cabinet, can hold many specific items (Fig. 10.2). Niche in the wall behind the WC would house the cleaning material, toilet brush, etc. niches or shelves can also be created in the side wall of the bathing area or the bathtub which can hold toiletries and other accessories.

3. The lower half of a tall window can also be turned to storage advantage. In bathrooms that are provided with glass louvres, the lower half of the window can be blocked by fixing the plate glass for privacy, and build glass shelves across. The light will not be reduced much while the louvres in upper half with the addition of an exhaust fan would keep the bathroom well ventilated (Fig. 10.3). Plants in this area would brighten up the whole room.

4. The geyser can be housed in a loft above the bathing area which could also provide long-term storage for other bulky items that are impervious to moisture and damp. The water tank can also be placed in this loft. A door would hide everything neatly.

For the bathrooms with sufficient space for storage, need to be well planned to achieve the efficient design. If the bathroom has a counter basin, two types of storage can be provided.

Base Cabinet

The area under the counter allows for the drawers and cupboards. Apart from storing things, cabinet will conceal the trap and the pipes of the washbasin, giving it a neat look. The central portion of the cabinet, i.e. under the trap can be used for keeping cleaning material, etc. The side

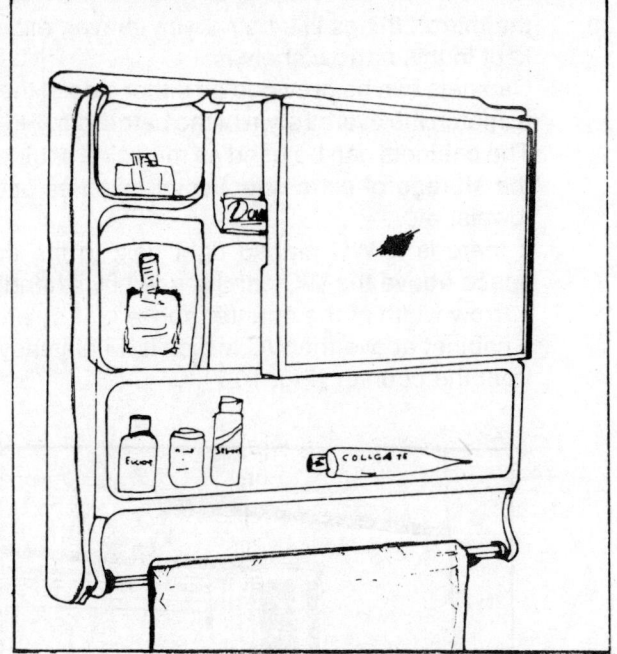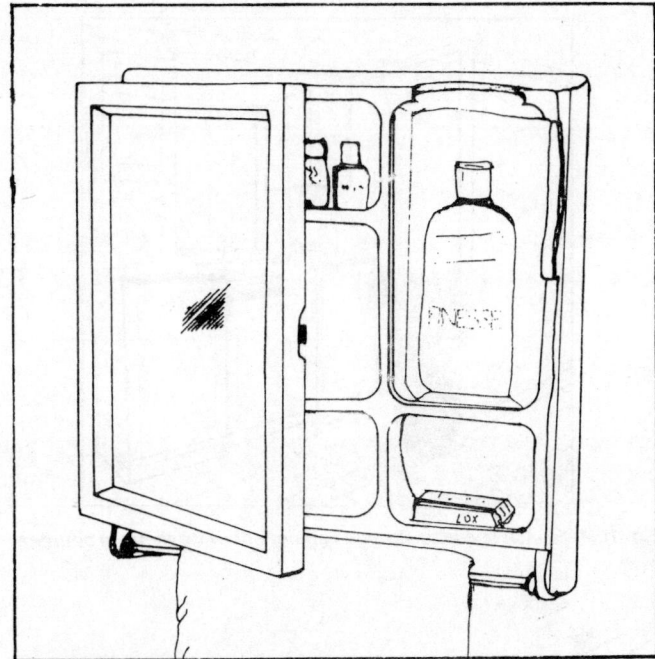

Fig. 10.1 View of the storage cabinet with sliding mirror door generally used over washbasins for storage of toiletries.

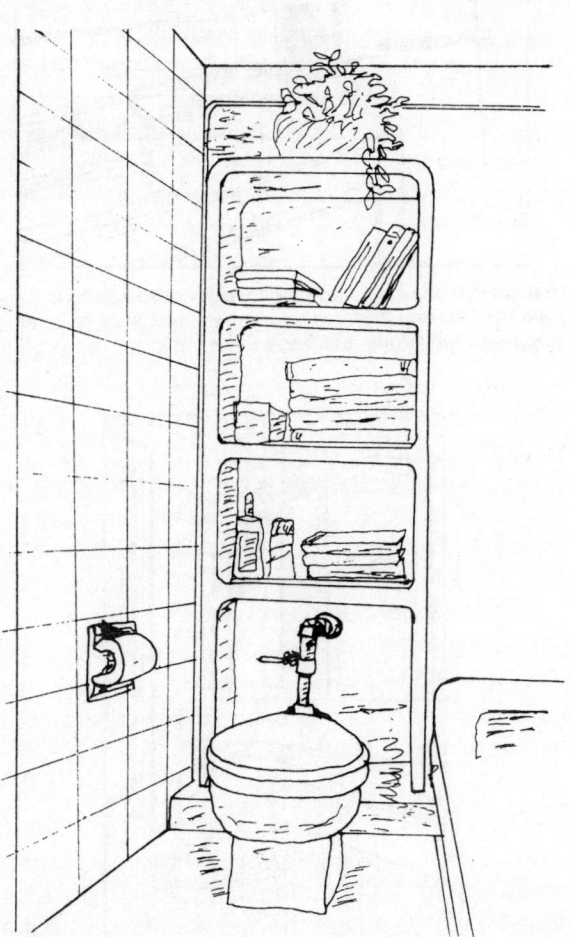

Fig. 10.2 Sketch showing niches or shelves in the walls in addition to the cabinets, etc.

Fig. 10.3 Glass shelves in the lower half of a window furnish a simple and practical storage solution.

cabinets can be used for other purposes, like dirty linen, clothes, extra supplies of toiletries, undergarments, towels, etc. If the dressing area is also included in the bathroom, the place for cosmetics, hair dryer, etc. can be created by providing top drawers in the cabinet. The cabinets should be kept off the floor to facilitate easy cleaning as well as to prevent it from the dampness (Fig. 10.4).

Storage above the Counter

The space can be created for the toiletries which are used regularly on the counter itself. Smaller items of bathroom storage are more functional than the size implies. Toothbrush racks, mug holders and soap dishes all help to clear the decks. Some small cabinets or shelves around the mirror with or without door for medicines or some miscellaneous items to be needed while using the washbasin (Fig. 10.5). Following are some of the alternatives which can be used for storage above the counter.

Fig. 10.4 *Sketch showing storage cabinet below wash basin counter.*

Fig. 10.5 *Sketch showing storage above wash basin counter.*

the mirror, things like hair dryer, shaver, etc. can be kept in this narrow shelves.

2. Cabinets can be provided on either side of the mirror to match the overall style of the bathroom (Fig. 10.6). The cabinets can be used as medicine cabinet or for the storage of extra supplies of toiletries and hand towels, etc.

3. If there is a WC placed right next to the counter, space above the WC can be used by extending the narrow width of the counter above it or by providing a cabinet above the WC, which is within easy reach from the counter (Fig. 10.7).

Fig. 10.6 *Sketch showing storage cabinets above washbasin counters on the sides of looking mirror for storage of medicines or for storage of extra supplies of toiletries and hand towels, etc.*

Fig. 10.7 *Counter space at the back of water closet being used as extension of washbasin counter.*

1. The counter basin usually have walls on either end and mirror on the back wall. An alternative is to provide shelves on the side walls with or without a door, hinged at the mirror end. Since, the electrical points are near

Closets in Dressing Areas

Modern bathrooms generally have attached dressing areas. The clothes storage in dressing areas is very crucial and needs careful planning, so that all the things can be nearly stored and at the same time, have sufficient clearances. Much can be stored in little space if sufficient thought is given to the equipment. Too many closets have unused and unusable space due to poor planning. Good closet design requires planning, arrangement, and fixtures contributing to:

1. Convenience
   a. ease of access
   b. maximum visibility
   c. orderliness
   d. maximum accessibility
   e. maximum of used space
2. Preservation
   a. of pressed conditions
   b. of freshness (ventilation)
   c. from moths
   d. from dust
   e. from pilfering

Doors of the closet should open full width of the closet whenever possible. In most cases, the most efficient and economical doors are the usual hinged type. Two doors for 1.5 closet will eliminate dark, inaccessible, hard-to-clean corners. Hooks, racks, and accessories on the backs of swinging doors increase efficiency by using otherwise unoccupied space in the closet.

In case of small areas, sliding doors may be used to avoid the obstruction in the room. Sliding doors can expose the entire interior of the closet to view and make it immediately accessible. Such doors do not block traffic. Sliding doors, however, do not permit the use of special fixtures such as tie racks, shoe racks or mirrors, which are handy and easily reached when attached to a hinged closet doors (Fig. 10.8).

The standard depth of the closet used for hanging clothes, is 600" 750" (if a hook strip is used). This permits clothing to be on hangers on rods, with sufficient clearance. Clothes closet width, parallel to the doors, should be from 3 to 6 ft. per person, depending on amount of clothing and whether drawers are to be provided (Fig. 10.9).

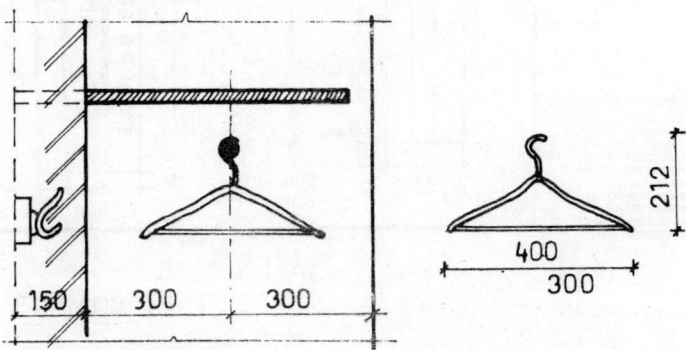

Fig. 10.9 *Section showing minimum effective depth of closets and hanger sizes.*

Basic Elements

The standard elements of closet storage are shelves, drawers, rods, hooks and special fixtures (Figs 10.10A to C).

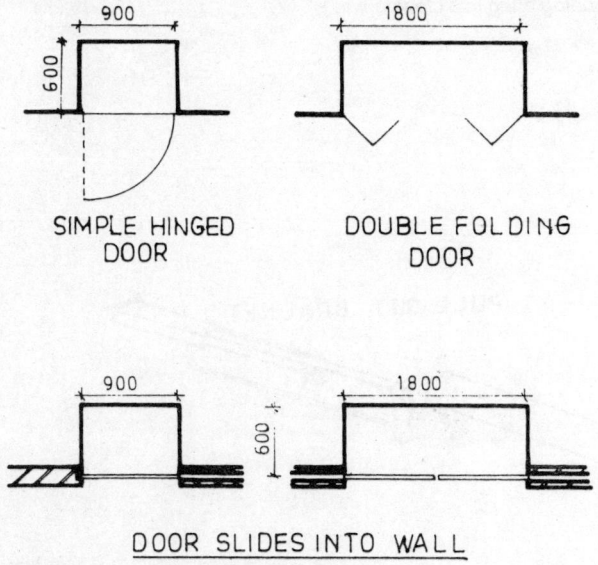

SIMPLE HINGED DOOR

DOUBLE FOLDING DOOR

DOOR SLIDES INTO WALL

DOOR SLIDES BEHIND EACH OTHER ONE HALF OF CLOSET MAY BE OPENED

Fig. 10.8 *Plan showing single hinged door, double folding hinged door for wardrobes/dressing closets.*

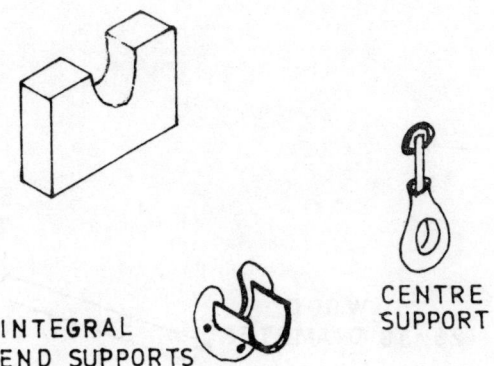

INTEGRAL END SUPPORTS

CENTRE SUPPORT

Fig. 10.10A *Various types of end and central support for closet rods.*

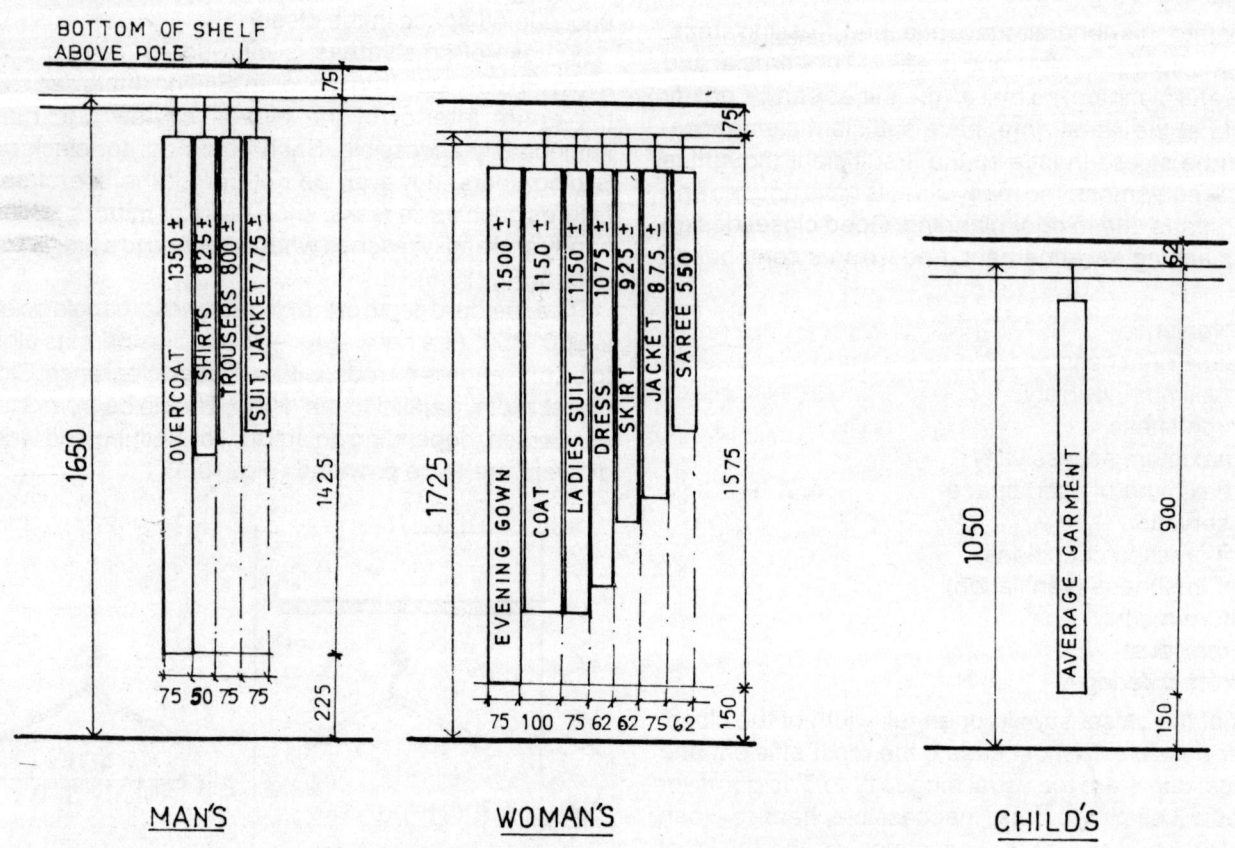

BOTTOM OF SHELF
ABOVE POLE

MAN'S

- OVERCOAT 1350
- SHIRTS 825
- TROUSERS 800
- SUIT, JACKET 775

1650 · 1425 · 225 · 75 · 50 · 75 · 75

WOMAN'S

- EVENING GOWN 1500
- COAT 1150
- LADIES SUIT 1150
- DRESS 1075
- SKIRT 925
- JACKET 875
- SAREE 550

1725 · 1575 · 150 · 75 · 100 · 75 · 62 · 62 · 75 · 62

CHILD'S

- AVERAGE GARMENT

1050 · 900 · 150 · 62

Fig. 10.10B *Details of sizes of clothes being hung in a closet.*

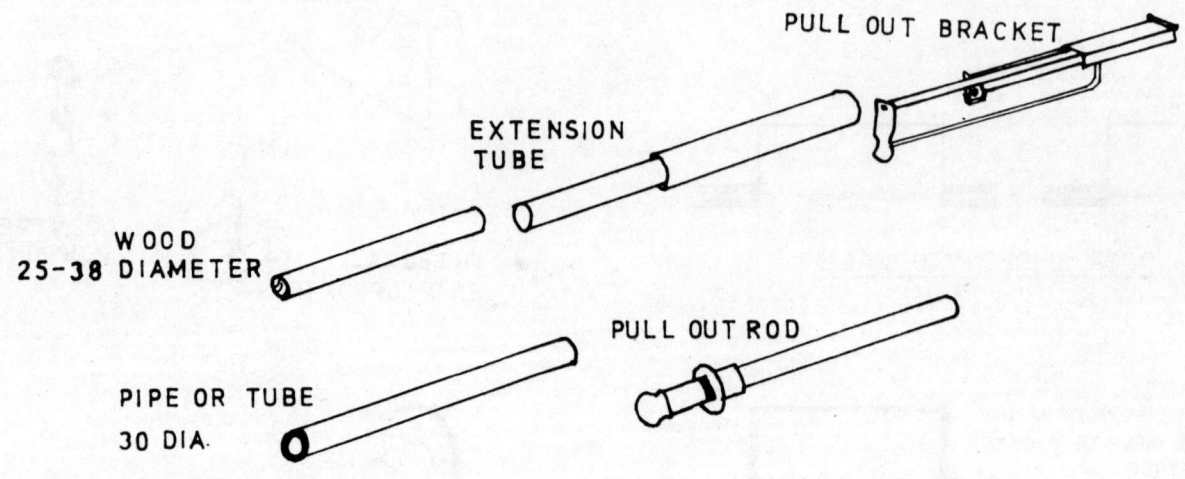

PULL OUT BRACKET

EXTENSION TUBE

WOOD 25-38 DIAMETER

PIPE OR TUBE 30 DIA.

PULL OUT ROD

Fig. 10.10C *Various type of hanging rods.*

Shelves

Shelves are simple and inexpensive to install, require a minimum of effort to use, and are adaptable to storage of many types of things, especially those of odd or bulky shape, folded articles and books, magazines, etc. A 300 mm self is usually adequate for most things. Articles of large dimensions or greater depth should have their special places, linens, for instance, are frequently folded for a 450 mm, shelf. Some useful shelf-support methods are depicted in (Fig. 10.11).

Drawers

Drawers can very conveniently accommodate various things with a minimum of space. They present a neat appearance even when carelessly used. Drawers of different widths and depths make possible filling of various known possessions of the user, providing a great saving in time.

Drawer construction requires both skilful craftsmanship and best materials. They must operate freely under all climatic conditions. Figure 10.11 shows various drawer constructions.

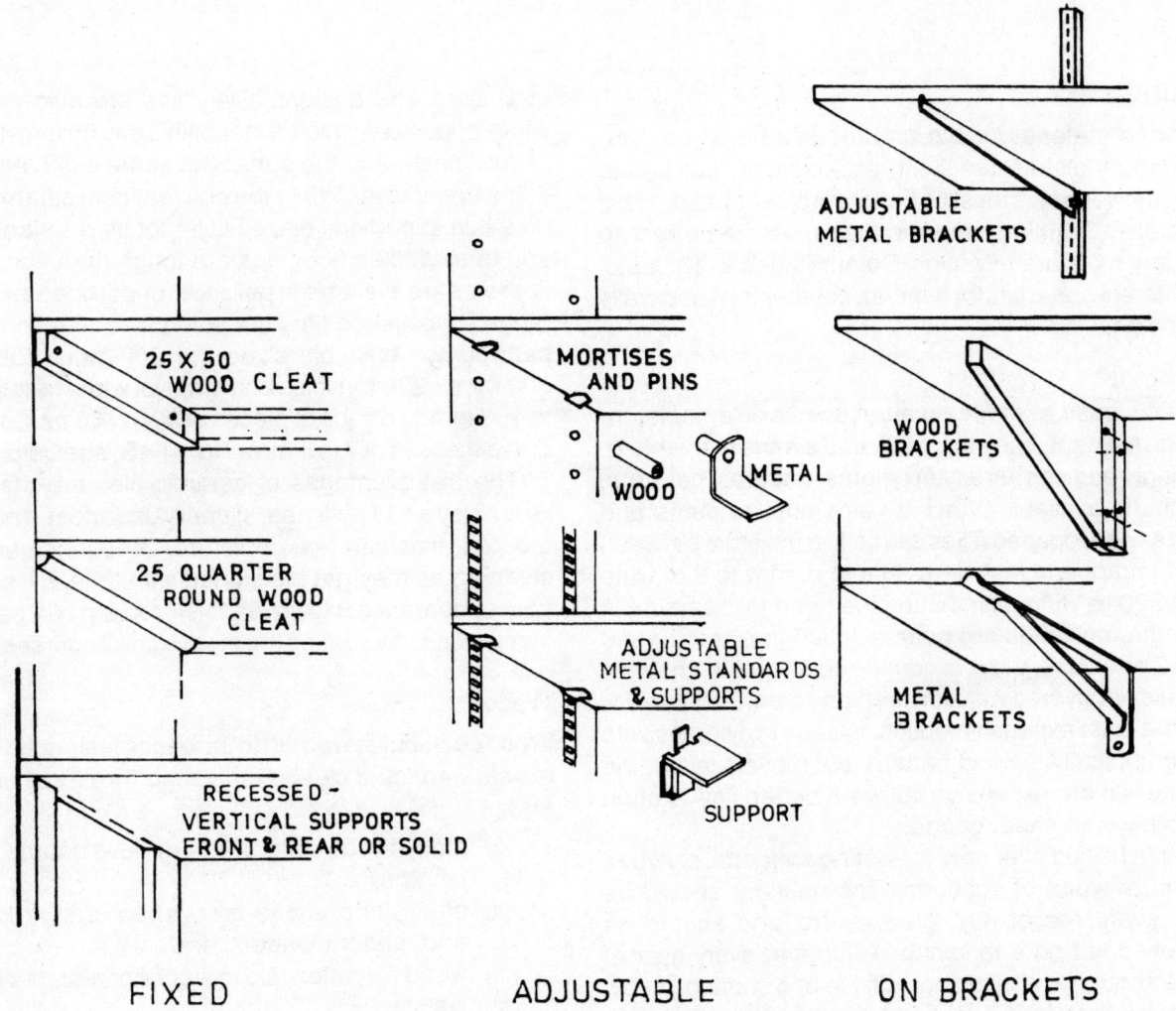

Fig. 10.11 *Various details of shelf support methods.*

MATERIAL APPLICATION

FLOORING

The flooring material for the bathrooms should be non-slippery and moisture resistant, especially in wet areas. Grouting between tiles must be perfect to prevent, especially in wet areas. Grouting between tiles must be perfect to prevent any moisture retention. Colour Plates 9–16 depict use of different materials for flooring, counter tops and walls in bathrooms.

Resilient Vinyl

A synthetic, easy to maintain attractive flexible, moisture and stain-resistant, easy to install, and simple to maintain. A seemingly endless variety of colours, textures, patterns, and styles is available. Vinyl is vulnerable to dents and tears but can be repaired. Tiles can collect moisture between seams if improperly installed. Sheets run up to 3 m wide and 16 to 20 m, long, eliminating the need for seaming in many bathrooms, tiles are generally 300 mm square and 2.0 to 3.2 mm thick. Vinyl is comfortable to walk on. Vinyl tiles should be covered with polyurethane paint for durability. Vinyl is the least expensive option. Some vinyl comes with a photographically applied pattern, but most is inlaid, the latter style is more expensive but wear better. This is often easier to lay than sheet goods.

Tiles can be laid over new or existing concrete, plywood or any other types of subfloors. The subfloor should be smooth, even free of dirt, grease, etc. and should be completely dried prior to laying. A thin and even layer of adhesive should be applied on both tile and subfloor. Allow the adhesive to touch dry before fixing the floor. Press the tiles firmly on the prepared subfloor. Vinyl flooring is one of the options for the dry area of the bathroom and should not be provided in the wet area or shower area of the bathroom (Fig. 11.1A, see Colour Plate 9).

Ceramic Tile

These are made of hard-fired slabs of clay, tiles are either unglazed (unfinished) red-clay tiles that are rough and water resistant, or glazed tile, available in glossy, matte, or textured finishes and in many colours. Ceramic tiles are available in various sizes and thicknesses ranging from 100 to 400 mm and 7 to 8 mm, respectively. These are available in variety

of shapes and designs. Clay tiles are also available as vitrified floor tiles which is monolithic and uniform throughout its thickness, and the surface is same even after wearing of the upper layer. While the conventional ceramic tiles rely on a thin superficial glazed layer for its resistance to wear and tear, vitrified floor tiles are tough throughout. Vitrified floor tiles are available in polished, unpolished and industrial range. Unpolished tiles are suitable for the flooring of the bathrooms. Available sizes are 200 mm × 200 mm and 300 mm × 300 mm tiles and strips of width ranging from 50 to 150 mm and square pieces of 50 to 150 mm, square and in thicknesses of 7 to 8 mm (Fig. 11.1B, see Colour Plate 9).

The disadvantages of ceramic tiles are that they are very noisy and if polished, slippery underfoot. If not properly grouted, tiles can leak moisture. Groutings need regular cleaning as they get blackened with time, giving it a very ugly appearance. However, this problem can be solved to some extent, by sealing the grout with silicon-sealant paste.

Wood

Wood contributes warmth to the decor, feels good underfoot, resists wear, and can be refinished. The three basic types are:

  (i) strip, narrow tongue-and-groove boards in random lengths,
  (ii) plank, tongue-and-groove boards in various widths and random lengths, and
  (iii) wood tile, often laid in wocks or squares in parquet fashion.

Parquet flooring will give a good appearance to the bathroom floor. It consists of wooden blocks (teak, cedar, rubberwood and rosewood, etc.) fitted together to create geometric patterns on the floor as per design of choice. For its use in the bathrooms, it needs polyurethane finish on top.

The available parquet strips range from 8 to 20 mm thickness (25 × 8 × 150 mm) to (70 × 15 × 420 mm). (Fig. 11.1C, see Colour Plate 9).

Moisture will damage wood flooring, thus, wooden flooring is rarely used in bathrooms. Wood is moderate to expensive in cost, depending on quality, finish, and installation. The flooring should be suitably treated with polyurethane, varnish or paint, but it is susceptible to damage even if treated.

Wooden laminate flooring It is the recently available product in India. It can also be used in bathrooms but recommended only for dry area and is much more durable than other types of wooden flooring. This type of flooring is resistant to stain and moisture, can sustain wear and tear. Manufactured by direct pressure laminate technology, these laminates have hard, resin-based surface coating for durability. These panels are provided with tongue and groove, which are sealed to protect them against moisture and prevents swelling at the edges. These panels are available in variety of colours and designs to give various effects and are available in 8 mm thickness.

Stone

Natural stones such as slate, flagstone, marble, granite and limestone, have been used for flooring for centuries. Today, its use is even more practical, thanks to the development of sealers and finishes. Easy to maintain, stone flooring is also virtually indestructible. Stone can be used in its natural shape or cut into uniform pieces rectangular blocks or more formal tiles. Generally, uniform pieces are butted tightly together (Fig. 11.1D, *see* Colour Plate 9).

The cost of most stone flooring is high. Moreover, the weight of the materials requires a very strong, well supported sub-floor. The most commonly used stone for bathroom flooring is marble. Marble is available either in slabs or tiles. Slabs are available in 30 to 220 cm length, 30 to 100 cm width, and 2 to 15 cm depth. Tiles are available in square of 10 cm × 10 cm to 60 cm × 60 cm and thickness 1.8 × 2.5 cm. Marble slabs and tiles are available in the following finishes:

(i) Sand abrasive finish—a flat non-reflective surface,
(ii) None finish—a velvet smooth surface with little or no gloss.
(iii) Polished finish—a highly finished glossy surface which brings out the full colour and character of marble.

Terrazo These are granular marble chips laid in cement. This flooring requires very little maintenance. Terrazo is smooth and tough. It can be trowelled, rolled or laid down in the form of slabs or hydraulically pressed tiles.

It is basically non-slip, available in a range of colours, and washes clean with hot, soapy water. But the only disadvantage is that it tends to turn black near skirting.

Carpet Generally in cold and humid zone, the toilets are also provided with wall to wall carpet in the areas other than bathing area to suit the climatic requirements (Fig. 11.1E, *see* Colour Plate 9).

WALL COVERINGS

Bathroom wall treatment must be able to withstand moisture, heat and high usage. These surfaces play an important part in defining overall impact of the bath.

Ceramic Tiles

Wall tiles can be polished, and offer a great variety in colour and design. Generally, lighter and thinner than floor tiles, they are primarily on interior surface—walls, countertops, and ceilings. Their relatively light weight is a plus point for vertical installation. Standard sizes for wall tiles range from 100 mm squares to 400 × 300 mm rectangles, thickness vary from 6 to 9 mm, other sizes and shapes are also available. Many wall tiles come with matching trim pieces for edges, corners and borders to give decorative effect to the bathroom walls (Fig. 11.2A, *see* Colour Plate 10).

Stone

Stones such as marble, granite, limestone, etc. are good options for wall treatment. Normally, stone tiles are used for the walls and can perform a role similar to ceramic tile. Stone tiles are expensive but used as accents, they can go a long way (Fig. 11.2B, *see* Colour Plate 10).

Glass Blocks

If you are looking for some ambient daylight but do not want to sacrifice privacy, consider glass blocks. You can buy 8 to 10 cm thick square blocks in many sizes and shapes. Textures can be smooth, wavy, ripped, bubbly or cross-hatched. Mortarless block systems make an often tricky installation job. The mortar joints of glass block units act as louvres to regulate the sun's heat and light. By product and pattern selection, one can control lighting with diffusion, reduction or reflection. Glass blocks are available in various sizes ranging from 15 to 24 cm square generally used for shower enclosures (Fig. 11.2C, *see* Colour Plate 10).

Wall Paper

Wall paper for the bathroom should be scrubbable, durable and stain resistant. Solid vinyl wall papers, which come in a wide variety of colours and textures fit the bill. New patterns including some that replicate others surface (such as lines), are generally subtle. Wall paper borders add visual punch to ceiling lines and openings. Good ventilation is crucial to keep wall paper from loosening (Fig. 11.3A, *see* Colour Plate 10).

Wood

Tongue and groove wood panelling—natural, stained, bleached or painted—provides a charming accent to country schemes. Wood laminated particle boards or block boards can be used for all panelling at the places where wood panels do not come in direct contact of water. These laminates are manufactured by timber chips or blocks bonded with phenol formaldehyde synthetic resin in different layers and then laminated by hot pressing melamine papers on both sides under high temperature and pressure. In bathrooms, laminates should not be flushed with wall but in the form of panels with wooden frames. For durability, these boards are lined with hardwood strips or at places where it can come in contact with water. Any type of wood can be used in the bathroom, provided the bathroom is well ventilated (Fig. 11.3B, *see* Colour Plate 10).

These tiles give a very colourful effect and can be used on one or two walls to give a special effect to the bathroom. These tiles can also be used as a wall mural. The size of one glass mosaic tiles is 20 mm, 20 mm and a thickness of 4 mm. These tiles are available in sheets of 320 mm, 320 mm.

MOULDINGS AND CORNICES

Various mouldings and cornices are available in plaster of Paris wood or stone to give decorative effect to the walls and ceilings of the bathrooms (Fig. 11.4A). These can be used at various places in the bathrooms.

   (i) borders to match the wall tiles
   (ii) around windows or mirrors
   (iii) corners of the ceilings
   (iv) to highlight any other feature or part.

(Fig. 11.4B, *see* Colour Plate 11).

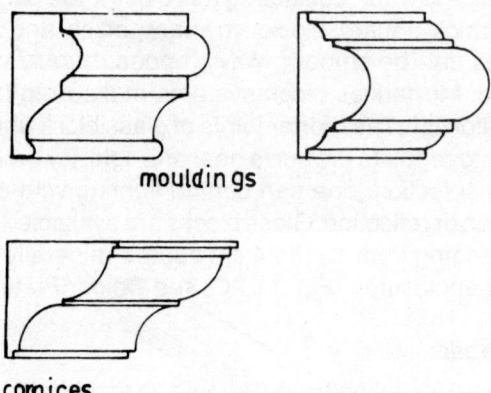

mouldings

cornices

Fig. 11.4A *Various types of POP/wood moulding and cornices.*

COUNTERTOPS

Stone is the major countertop material currently in use. Solid surface acrylic, ceramic tile, plastic laminate and synthetic marble are also quite popular. Wood can also be used for countertops, but the surface must be carefully sealed to prevent water damage.

Laminates

Laminate comes in a wide range of colours, textures, and patterns. Its easy to clean, water-resistant, and relatively inexpensive. However, it can scratch, chip, and stain. Also, smooth, reflective surfaces tend to show dirt and water marks. Conventional laminate has a dark backing that shows at its seems—new solid colour laminate, designed to avoid this, is somewhat brittle and more expensive.

Post formed tops, premoulded and prefabricated, are the least expensive option, a custom top with built-up lip and backsplash looks best but is more costly (Fig. 11.5, *see* Colour Plate 11).

Ceramic Tiles

Good looking ceramic tiles are available in many colours, textures and patterns (Colour Plates 12 and 13). Installed correctly, it is heat proof, scratch resistant, and water resistant. For durability, the joints between the ceramic tiles should be properly grouted.

Prices range from modest to extravagant, depending on style, accents, and accessory pieces. Nonporous glazed tiles do not soak up spills and stains. The tiles are available in sizes ranging from 200 to 400 mm, in square and rectangular shapes (Fig. 11.6, *see* Colour Plate 11).

Stone

Granite and marble are beautiful natural material for countertops. In most areas, one will find great selection of colours and figures. Stone is water resistant, heat proof, easy to clean and very durable. Oil, alcohol, and any acid (even chemicals in water supplies) can stain marble and damage its high gloss finish, granite can stand all these. Solid stone slabs are very expensive. Stone tiles, including slate and limestone, are less expensive alternatives. Edge mouldings in stone are also available to give a neat finish to the edges of the countertops (Fig. 11.7, *see* Colour Plate 11).

Cast Polymers

This group of man-made products, collectively known as "cast polymers', includes cultured marble, cultured onyx, and cultured granite. All three are relatively inexpensive and easy to clean. These products are often sold with an integral sink. Synthetic marble is not very durable, and scratches and dings are hard to mend (the surface finish is only a thin veneer). It is available in the market in a wide range as polycast marble (Fig. 11.8, *see* Colour Plate 11).

Wood

Wood is handsome, natural, easily installed, and easy on grooming accessories and glass containers. However, wood is harder to keep clean than nonporous materials. If you use it in areas that will get wet (and that includes many bathroom surfaces), thoroughly protect it on all sides with a good sealer, such as polyurethane. Wooden laminated particle boards and block boards can also be used as countertops, but the edges of the boards should be properly lined with either hardwood or edge mouldings (Fig. 11.9, *see* Colour Plate 11).

HOW TO LAY CERAMIC TILES
Fixing Wall Tiles

Wall tiling a whole room involves a large scale tiling. Sometimes, you will have to cut and shape tiles to clear obstructions and awkward corners. Apart from the tools needed for the tiling, you will need the following tile accessories.

Plastic Edging and Sealing Strips

Plastic edging gives a neat finish to corners and edges where the tile edges are unglazed. Sealing strips forms a watertight seal along the joints with worktops, bathtubs and basins. Both types are simply bedded in the tile adhesive before fixing the last row of tiles. They are sold in various lengths in a range of colours (Figs 11.10A and B).

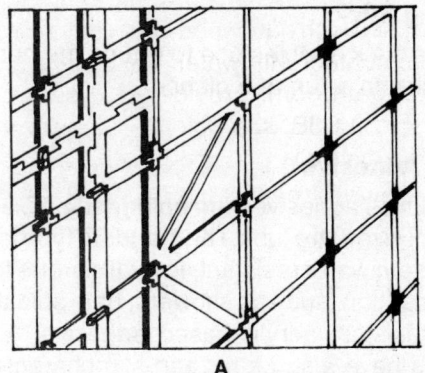

A

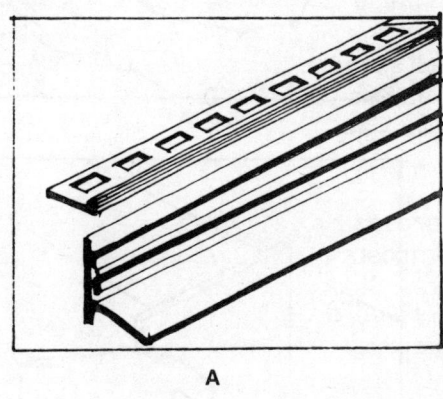

A

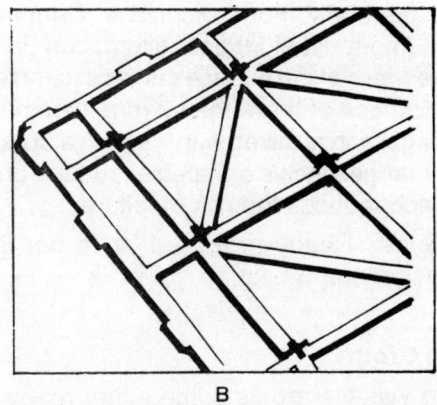

B

Figs 11.11A and B Details of trutile tiling grid.

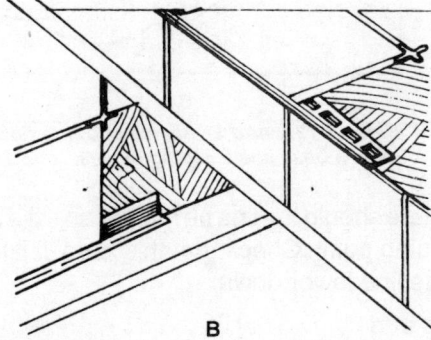

B

Figs 11.10A and B Details of plastic edging and sealing strips.

Trutile Tiling Grid

Ingenious system of plastic interlocking grids which are stuck to the wall with tiling adhesive or household glue. Small cross-pieces within the grids then allow tiles to be spaced and levelled automatically. Obstructions are easily cut around with scissors or a knife does away with the need for traditional setting out with battens, but can work out expensive on large areas and only available for certain tile sizes (Figs 11.11A and B).

Flexible Sealant

Silicon-or acrylic-based sealant used to fill gaps up to 6 mm wide between tiles and worktops or plumbing fixtures (sealant slump in wider gaps—use quadrant tiles or sealing strip instead). Also used in place of grout on tiled panels that need to be removed (Figs 11.12A and B). These are bought in cartridge form.

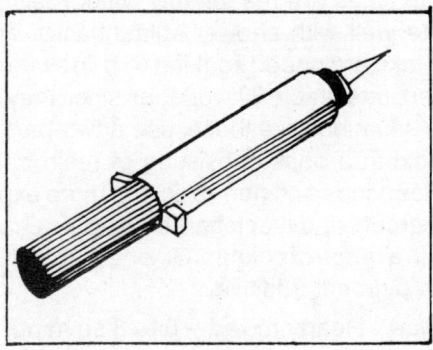

A

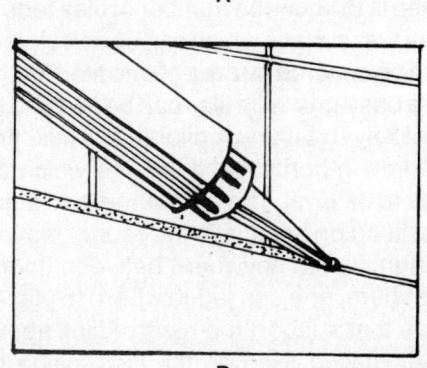

B

Figs 11.12A and B Sketch showing application of flexible sealant with the help of cartridge.

A Marking Stick

A marking stick enables one to gauge the number of tiles and cut tiles in a run at a glance.

Wall Tile Adhesive

Most wall tiles adhesives are sold ready mixed in a range of different sized litre tubs. The standard type is PVA-based and only semiwater resistant, for walls prone to dampness or condensation (such as showers, bath splashbacks) use a water-resistant acrylic-based adhesive. Acrylic-based adhesives have also better non-slip characteristics, and some types can be used to grout the tiles as well, however, they are generally more expensive. Cement-based tile adhesive (more usual for floor tiling), can be used where the unevenness of the surface makes it necessary to apply a bed thickness of more than 3 mm. It comes in powder form in bags, and is mixed with water in a bucket. Many tile adhesive ranges include a special surface primer, sold in cans, which reduces the risk of failure.

Cover guide Ready mixed—1 sq m per litre, powder form—1 sq m per 3.5 kg.

Wall Tile Grout

Standard wall tile grouts come either ready mixed or in powder form. Ready mixed grouts are acrylic based and sold in tubs. Powder grouts are cement based, and come in bags for mixing with water, they are slightly easier to apply. Both types are reasonably water resistant and can cope quite well with shower splashbacks. They cannot, however, take prolonged soaking (e.g. in swimming pools), and they are unsuitable for worktops since they can harbour germs. In situations like these, use a two-part epoxy resin grout—sold in a pack consisting of resin and hardener. This is impervious and non-toxic, but more expensive than standard grouts and much harder to apply. Grouts are now available in a range of colours. Or, one can colour powdered grout with pigment additive.

Cover guide Ready mixed—6 to 8 sq m per kg, cement based—1 sq m per kg.

First step is to know the number of tiles required to cover a particular wall surface which can be calculated by dividing the area to be covered by area of one tile. For the wall tiling, you need a base on which tiles can be laid and the floor and skirting is likely to be uneven for that. Thus, first of all one needs to draw a horizontal base line which allows every row of tiles to be level. The next stage is to adjust the base line, depending on what is in the room, to avoid having to make unsightly cuts anywhere between floor and ceiling level. From here, one can judge where to place the setting out battens that support the rows of tiles above and keep them level. Having fixed all the tiles inside the battens, remove the battens and fill in the remaining gaps. The base line can be drawn by following procedure.

Using a batten and spirit level, draw a level line right around the room just under a tiles width above where the tiles have to finish (Figs 11.13A and B).

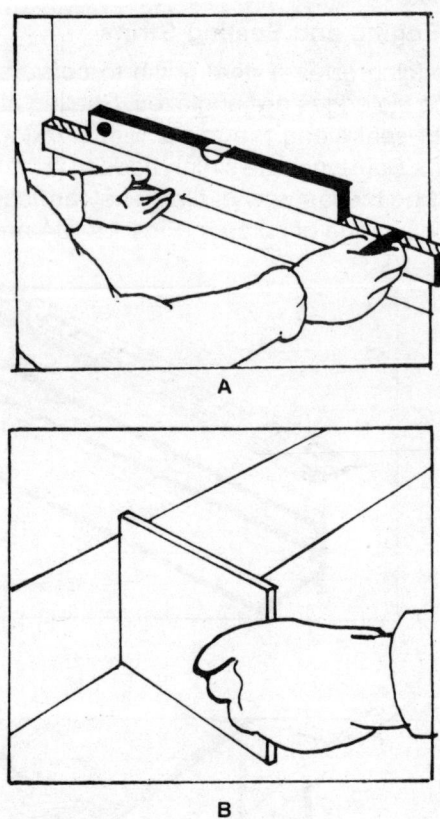

Figs 11.13A and B *Sketch showing how to draw a base line for fixing of tiles.*

Nowhere should the line be more than a tile's width above the finishing point. Check that this is so, if it is not, draw a new base line lower down.

SETTING OUT THE ROOM

Setting out allows you to place cut tiles where they will be least noticeable, and shows where to fix the support battens. Start by using marking stick to measure the tile widths between the base line and any fixtures on the wall. Follow the sequence shown on the right (Fig. 11.14), and mark where the cuts fall. Then adjust the height of the base line to get rid of cuts where you do not need them—e.g. along a window sill, or the top edge of the bath panel. After doing this, redraw the base line right around the room. This shows where to fit the horizontal support battens.

Use the same technique to check from side to side. Mark out each wall so that you avoid cut tiles at external corners and at the sides of windows. Where cuts are required at both ends of a wall, find its midpoint and measure out from here so that both lots are equal.

Finally, mark where the last column of whole tiles finishes on the left hand side of each wall. Plumb lines here, showing where to fix the vertical battens.

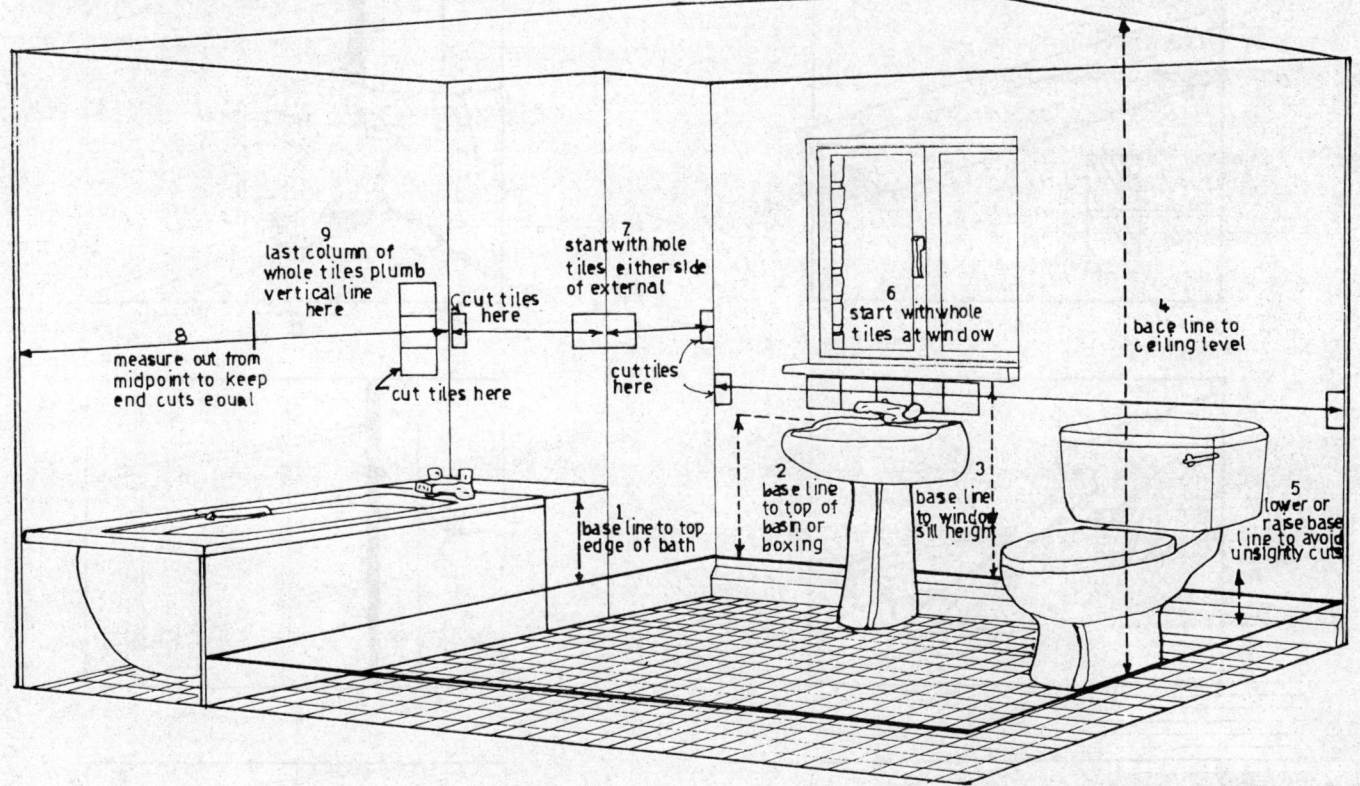

Fig. 11.14 *Various points of consideration while fixing tiles in a bathroom.*

Tiling with Battens

After making base line, part-drive nails into the first batten at 300 m intervals; the points should just show through (Fig. 11.15A).

Offer up the batten level with the base line and drive in the nails until they hold, leave the heads protruding so you can remove the batten later (Fig. 11.15B).

Having evened up the cuts from side to side, draw a vertical line to mark the last column of whole tiles. Fix the side batten against this line (Fig. 11.15C).

Spread about 1 sq m of adhesive in the usual way and fix the first tile in the corner of the two battens (Fig. 11.15D). Continue, working along the wall.

After completing the area inside the battens, leave it to set for an hour. Then slide a knife blade along the battens' edges to clear the joints (Fig. 11.15E).

Remove the battens by pulling out the fixing nails with pliers. Measure and cut tiles to fit gaps, and fix them in place in the usual way (Fig. 11.15F).

In case cement is used instead of chemical adhesive, the following procedure should be followed for ceramic tiles.

- Before fixing, the tiles must be soaked in water for about an hour.
- The mortar bed or surface in which tiles are to be fixed should be fine, smooth and free from dirt/dust particles to avoid contamination.

- Before fixing, tiles should be coated with thin layer of cement slurry sufficient to cover grooves to have better grip and strength and also to avoid air pockets.
- The mortar is to be spread on the back side of the tile. This should be composed on one part cement and two part sand.
- Pure cement paste must not be used to the back of tiles otherwise it may lead to cracks on the file at any later date.
- Tiles so placed of the mortar bed should be gently tapped to ensure total mortar contact.
- The setting of joints should be carried out minimum 24 hours after fixing the tiles.
- In order to obtain uniform tone, it is recommended to mix tiles from different boxes for laying.

FIXING FLOOR TILES

Floor tiling a room involves careful planning. Sometimes, the tiles are cut and shaped to clear obstructions and fixtures. Apart from the tools needed for the tiling, following tile accessories are needed.

Tile Accessories

Apart from the accessories used for the wall tiling, like tile adhesive, the grout, making stick, flexible sealant, etc. spacers and expansion joints are used for floor tiling.

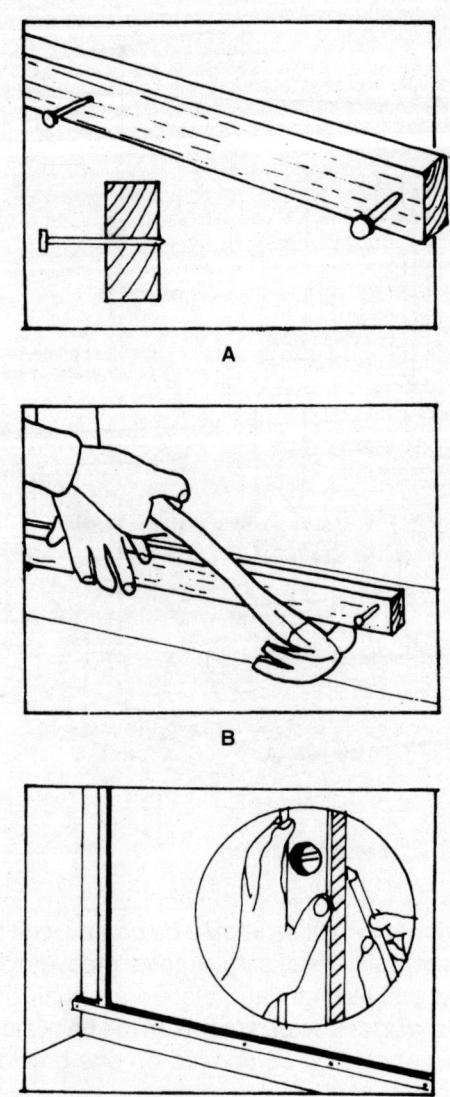

Figs 11.15A, B, and C *Use of wooden batten as a guideline for fixing of tiles.*

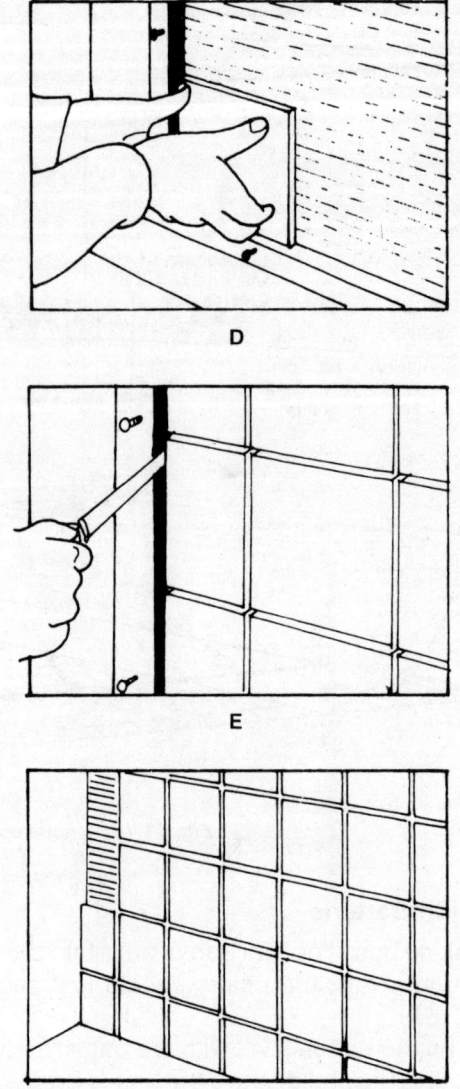

Figs 11.15D, E and F *Tiles being fixed having proper guide thumb points.*

Spacers and Expansion Joints

When cladding or covering any surface with ceramic tile or natural stone, the gaps between adjacent tiles play an important role. Technical function, as the inclusion of spaces helps to reduce the overall rigidity of the finished surface, which can be a cause of tension between the cladding and the subsurface. Aesthetic function, use of spacers will allow the gaps to accommodate any minor defects, size and rectangularity exhibited by the tiles. The spaces between the adjacent tiles are grouted later (Figs 11.16A to D).

Estimate and Setting out

First step is to know the number of tiles required to cover a particular floor which can be calculated by dividing the area (in sq m) to be covered by the area of one tile. To this, add 5 per cent to allow for breakage and wastage.

Then the layout of the tiles should be planned. Floor tiles can be laid out "dry"—i.e. without any adhesive—to check the layout, but for wall tiles, a gauge stick is used to even up tiles either side of a window. Gauge stick is a wooden stick sawn and levelled representing the spacing of tiles, used for setting out tiles.

Preparing the Surfaces

The two critical factors that determine perfect tiling are:

  (i) the surface to be tiled should be flat, and
  (ii) the surface should be sound, dry and clean of dirt and oil.

In case of area subject to excessive wet conditions and frequent water sponge, the surface should be treated for waterproofing before commencing fixing of tiles.

Figs 11.16A, B, C and D Sketches showing various types of spacers and expansion joints for tiles.

Preparing Floors for Tiling

Floor tiles need a dry flat surface and ideally concrete or existing mosaic tile floors when chemical adhesive is used. The surface on which tiles are to be fixed should be fine, smooth and free from dirt and dust particles to avoid contamination. The floor tiling always starts from the centre of the room.

Laying Ceramic Floor Tiles with Adhesive

Ceramic tiles can be laid on a bed of adhesive preferably by using plastic spacers to create a gap between tiles which can be grouted later on.

Lay around one square metre of adhesive at a time and press the tiles into place, with correct gap provided by spacers, available in different thickness (Figs 11.17).

Check that lines of tiles are straight. Allow 24 hours for the adhesive to set before cutting and fitting the edge tiles. The gaps between ceramic floor tiles are filled with matching grouts or contrast colours in the same way as wall tiles.

Wipe off the extra grouting (before it sets hard) with a damp sponge and polishing with a dry cloth.

In case cement is used instead of chemical adhesive, the following steps should be followed.

- Soak the ceramic tiles an hour prior to the flooring.
- 25 mm to 40 mm thick mortar bed in proportion 3:1 (sand:cement) by volume on the new floor and 12 mm on the existing cement floor or old mosaic floor.
- The slope, for water drainage should be provided in the mortar bed.
- Spread the screed with thin layer of cement slurry or chemical adhesive on the mortar bed to have better grip and strength and also to avoid air pockets.
- Place the tiles on the mortar bed and tap gently for total mortar contact.
- Bedding material between the joints should be cleaned with a brush before drying.

Grouting and Finishing

The gaps between ceramic tiles need to be filled with grout. Plain and coloured grouts are available to suit various tiles.

Leave the tiles to set for the time recommended by the adhesive manufacturers (normally at least 12 hours) before

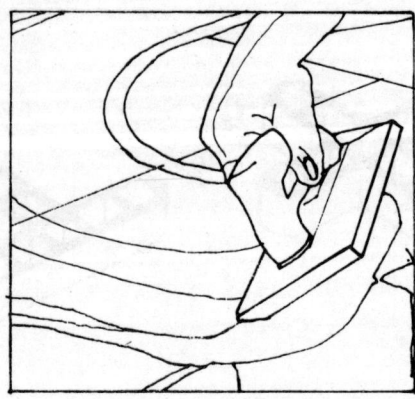

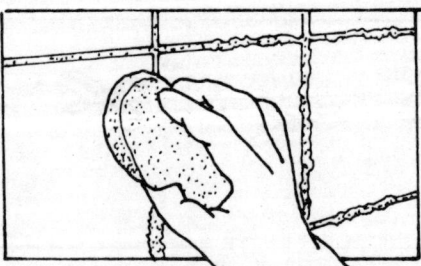

Fig. 11.18B *The excess grout should be wiped with a sponge washing it out frequently. No un-grouted pinholes should be left between the joints.*

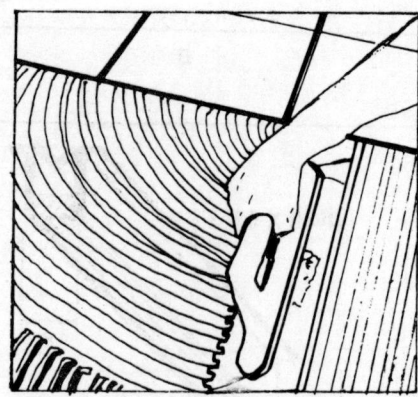

Fig. 11.18C *When the grout has begun to harden, rub down the joints to smooth them off to an even width.*

Fig. 11.17 *Tiles being laid on an existing PCC/mosaic floor with adhesive.*

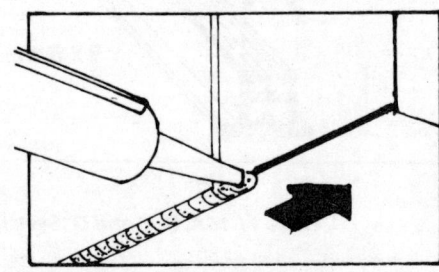

grouting. If the grout is in powder form, add it to the specified amount of water in a bucket and mix to a smooth, but fairly stiff, consistency. It is ready mixed, stir in the tub and then apply direct to the wall. Grouting is a very messy job, so cover everything else before starting. Grout hardens rapidly and becomes impossible to work into the joints, so make sure you have enough to do the job in one go. And, do not allow any grout to dry on the surface of the tiles, it is very difficult to remove one hard. When the grout has set, fill the joint with silicon or acrylic sealant. This provides a flexible seal that will not disturb the tiles if there is any movement. This silicon paste also prevents blackening of joints between tiles (Figs 11.18A to D).

Fig. 11.18D *Finally seal any gaps along the adjoining bath edge or work surface with sealant. The tube should be pushed away.*

Tiling Curved Areas

Objects like a pedestal basin or soil pipe have large radius curves that can be difficult to tile around. The answer is to make templates for each tile, then mark and cut the tiles individually to fit (Fig. 11.19A). In most cases, the object will be symmetrical so you can use the templates twice— once for each side. Do the actual cutting with a tile saw, add the extra adhesive when the cut piece is fitted (Fig. 11.19B).

Tiling Awkward Areas

(i) At *internal corners*, overlap one set of cut edges with another. Work out in advance which way to arrange the overlap so it is least noticeable (Fig. 11.19C).

(ii) *External corners* can be finished with trim strip. Bed the strip in the adhesive, then simultaneously fix both columns of tiles so that you can align them (Fig. 11.19D).

(iii) *Alternatively, overlap* tiles in the direction that is least noticeable—(Fig. 11.19E) around a bathtub, tiles on

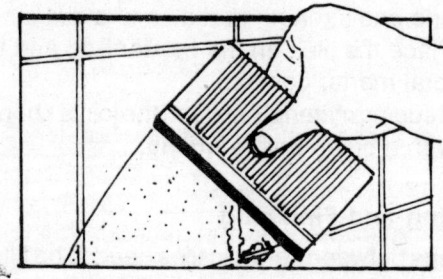

Fig. 11.18A *Adhesive spreader should be used to daub grout over tiles then quickly work it over the surface and into the joints with the grout spreader.*

a *horizontal surface* must overlap those on a vertical one).

(iv) Tile a *window recess* after the main wall. Arrange for equal size cuts on either side (Fig. 11.19F).

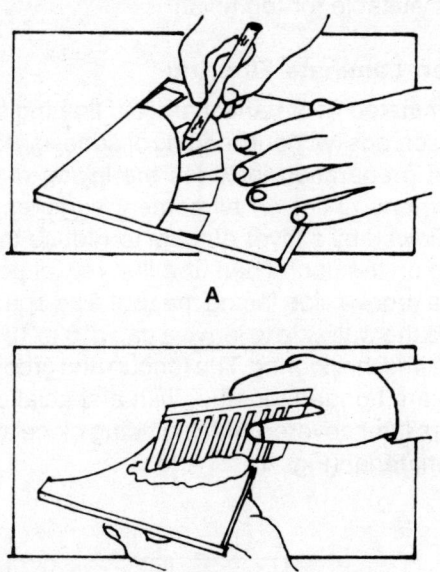

Figs 11.19A and B *Sketches showing how to cut and fix tiles in odd and curved places.*

Tiling Splashbacks

For a splashback or bath surround, if there is a firm level surface, such as worktop or bathtub edge, to use as a base. The aim should be to place the tiles where they look easiest on the eye, and to avoid unsightly cuts. Usually, this means finding the middle of the wall, and then tiling outwards from here so that any gaps at the ends are of the same width. However, narrow gaps look ugly, so the tiling should be started either on the midpoint or to one side of it. Do not forget to allow for 2 mm for the grout if the tiles are square edged.

For a basin splashback, mark the midpoint as shown on the previous page and then see which layout looks best by marking off in whole tile widths. Do not let the tiles overhang the edge of the basin too far—they will not have sufficient support.

Around a bathtub, plan things so that any cut tiles on the end walls are in the corner, allowing you to finish on whole tiles. If necessary, let the tiles overhang the tub slightly so that cut tiles are not too narrow. Having worked out the layout, draw a vertical line where the whole tiles end to use as a guide when fixing (Fig. 11.20).

HOW TO LAY MARBLE FLOORING

The marble should be laid in white or coloured mortar depending on the colour of the marble. The following is the procedure for laying of marble stone flooring.

1. First of all, prepare the base mortar consisting of OPC and fine aggregate (Bajri) in 1:5 to 1:6 proportion. The mortar should have low water to cement ratio.

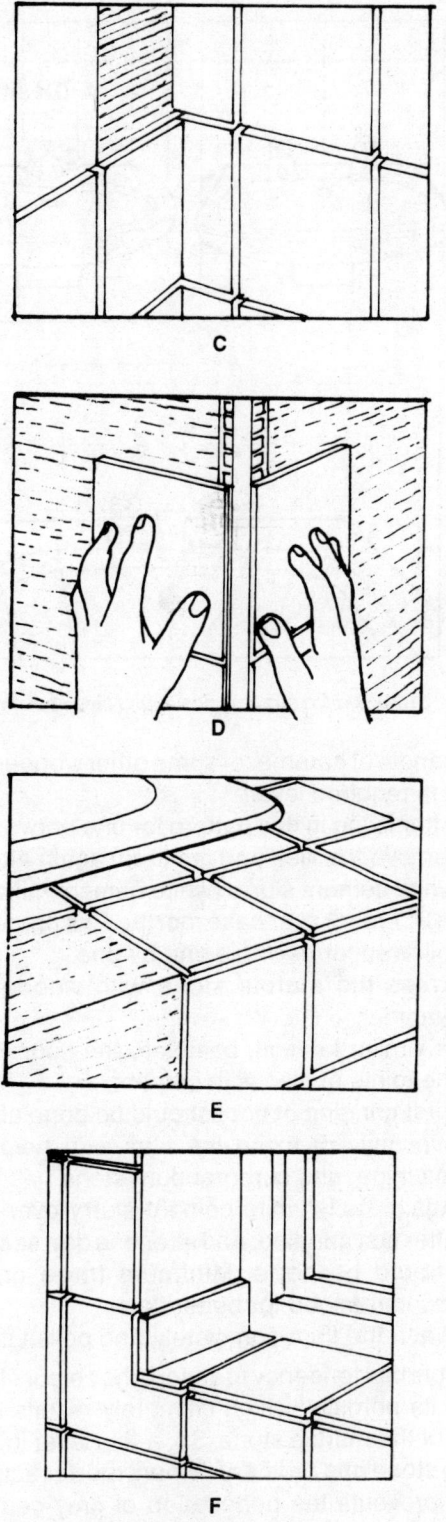

Figs 11.19C, D, E and F *Sketches showing fixing of tiles in internal and external corners.*

2. Minimum thickness recommended for base mortar is 25 mm.

3. Fix the marble tiles, temporarily with base mortar in the single or double row as per the design pattern by pressing it over the mortar with the help of wooden

Fig. 11.20 *Sketch showing position of tiles above wash basin.*

handle of hammer or some other wooden piece to get it in required level.

4. After fixing in this pattern for one or two rows of tiles, remove the tiles and put them again after spreading white cement slurry (white cement mixed with water in 2 : 1 ratio) over base mortar. This procedure is to be followed for each tile one by one.

5. Press the marble stone with wooden handle of hammer.

6. In similar fashion, complete the floor. Finally, fill up the joints of tiles with cement.

7. First grinding of floor should be done after minimum five days of fixing the tiles with help of grinding machine, and carborandum stone.

8. Again apply white cement slurry over marble floor after first grinding. And after one day, second grinding should be done. Minimum three grindings are recommended for better finish.

9. Wash the floor completely and polish it.

Marble has tendency to absorb the colour of base mortar due to its porosity, which ultimately results in change in colour of the marble stone. So, a thin layer in between the marble stone and ordinary cement mortar acts as a separator. It prevents the penetration of grey cement into the marble. This helps in retaining the original colour of marble.

HOW TO LAY WOODEN FLOORING

Parquet Flooring

Loose flooring strips are bounded directly as close as possible with parquet adhesive to the plain cement flooring or any hard level, smooth base, damp proof surface of existing floor. After fixing the strips as per pattern required 12 to 24 hours are required to dry the adhesive, after that it can be grinded and sanded with a dish fitted with different grits of sand paper which gives the floor the perfectly smooth surface suitable for top finish.

Wooden Laminate Flooring

The laminated floor panels are laid floating for the normal use on screeds (without bonding or adhesion to the surface). Surface preparation includes the laying of 0.2 mm thick polyethylene sheet on all cement surfaces and ceramic tiles followed by a layer of foam to reduce the noise while walking on the floor. Then, the first row of panels are laid, with the groove side facing the wall. The spacers are used towards the wall side to leave a gap of 8 to 10 mm between the wall and the flooring. The tongue and groove of adjacent panels are bonded together with a special glue. The side gaps are later covered with moulding piece available in the same material (Figs 11.21A to C).

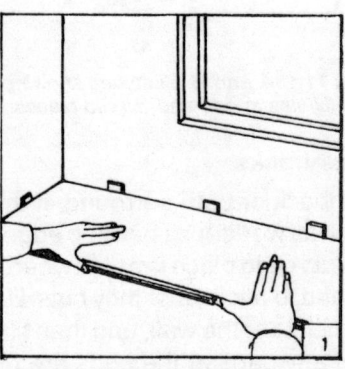

Fig. 11.21A *First row of panels are laid with the groove side facing the wall. No glue is used at this point. spacers are provided to leave a gap between the wall and the flooring.*

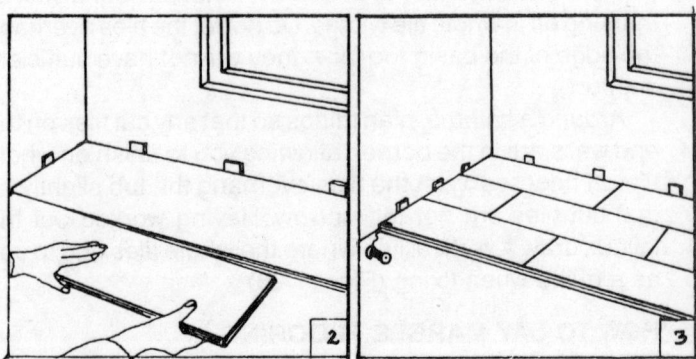

Fig. 11.21B *The last remaining piece of each row acts as the first piece of the subsequent row. The end joints should be spaced at least 20 cm apart.*

Fig. 11.21C *First two or three rows laid and checked with a cord to ensure that they are straight and at right angles. Then, they are bonded with glue with excess glue wiped with a moist cloth. The flooring should not be disturbed for two hours.*

HOW TO LAY GLASS MOSAIC TILES

The surface of all wall to be covered with glass mosaic tiles should be dry, clean and smooth. Spread the adhesive preferably rubber latex, on the wall by means of a toothed trowel of 2 mm (Fig. 11.22A).

Apply the mosaic sheets setting them in line so as to obtain a correct vertical and horizontal meeting of the joints. Tap the sheets with a wooden or rubber trowel to ensure adequate adhesion (Fig 11.22b).

The backing paper should be thoroughly mopped with a wet sponge. Wait for sometime so that the water seeps across and then peel the paper off, holding it by a corner. Fill the joints between the tiles with a slurry. Then clean the mosaic with a wet sponge and subsequently with a dry rag (Fig. 11.22C).

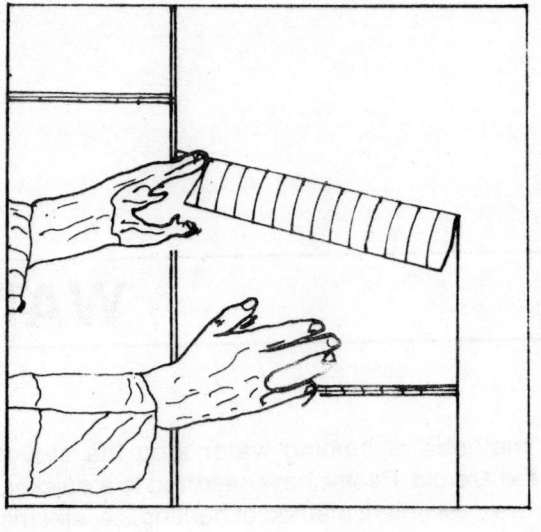

B

A

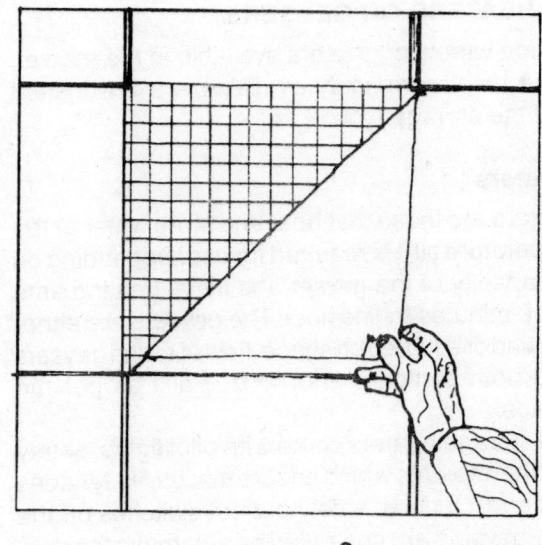

C

11.22A, B and C Sketches showing laying of glass mosaic tiles.

WATER HEATING SYSTEMS

Traditional methods of heating water atop the stove, immersion rod are out. People have resorted to a quicker, safer and more convenient method of heating, i.e. electric heaters or solar water heaters.

ELECTRIC HEATERS OR GEYSERS

There is a wide variety of geysers available in the market. Basically, the most commonly available are the instant geysers and the storage type.

Storage Heaters

Storage heaters are those that heat and store water at the desired temperature till it is required for use. Depending on the size or capacity of the geyser, the initial heating time varies from 10 minutes to one hour. The geysers are either the regular round or squared off shape. Few of round geysers can be fixed both vertically and horizontally as per the available space.

Almost all storage heaters come with pilot lights, safety cut-outs and thermostats which ensure that the water does not overheat and if it does, safety cut-out switches off the supply. Some geysers are controlled by automatic thermostats which switch off and on when the temperature rises above or falls below the desired warmth. The geysers are available in different capacities from 15 to 100 litres in the market. Available in single or two toned colours, in a vast variety of shades to match with the todays' distinctively designed bathrooms.

The storage water heaters should be operated only when they are filled with water. The geysers should always be fitted with pressure release valve or vent pipe and a nonreturn valve on the inlet water line. Never leave the top vent open. If not covered by vent pipe connection/dead weight type pressure release valve, plug top vent with fusible plug. Use pressure reducing valve for high rise and multistoreyed buildings.

Figures 12.1A to D show the general plumbing arrangement in pressure type installation. This type of installation is done where hot water is required at more than one place, e.g. shower, washbasin, bathtub, etc. The cold water supply must be taken from an overhead tank

and should not be connected directly to municipal mains. It is advisable to have a direct and separate cold water pipe line from the overhead tank to the inlet of the storage water heaters. If this is not possible, and tapping from the common downtake pipe is to be taken, then, it is necessary.

The Instant Heaters

The instant water heater gives off hot water immediately after switching on and it does not store water. The instant water heater has an outlet and an inlet. The inlet is directly connected to the cold water supply line. It does not require any plumbing. The instant geysers are available with the water mixers to get the water at required temperature.

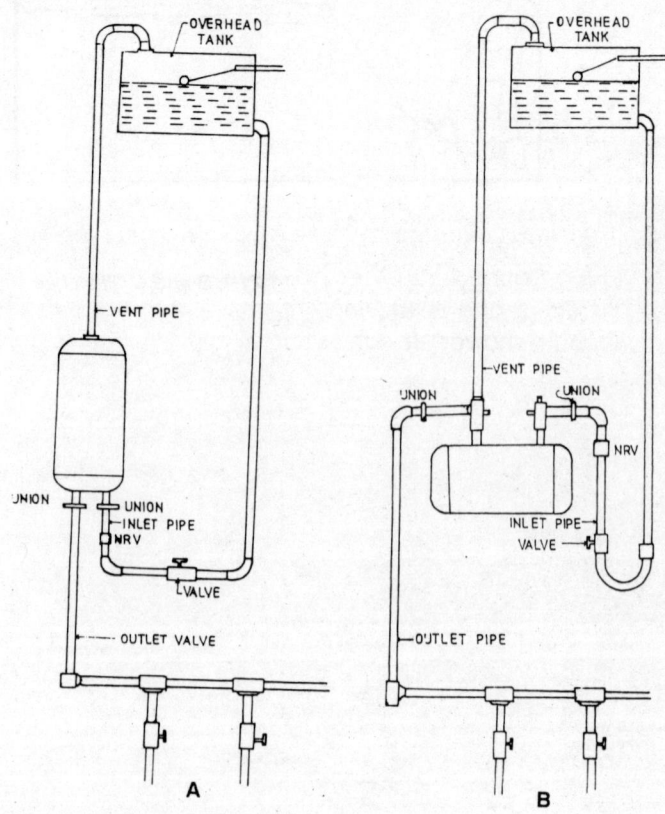

Figs 12.1A and B Sketches showing pressure type installation of geysers using vent pipes.

- AVV — ANTI-VACUUMVALVE
- PRV — PRESSURE REDUCINGVALVE
- NRV — NON-RETURNVALVE
- D P — DRAIN PLUG

- THE HORIZONTAL STORAGE WATER HEATER MUST ALWAYS BE SO MOUNTED THAT INLET AND OUTLET PIPES POINT VERTICALLY UPWARD.

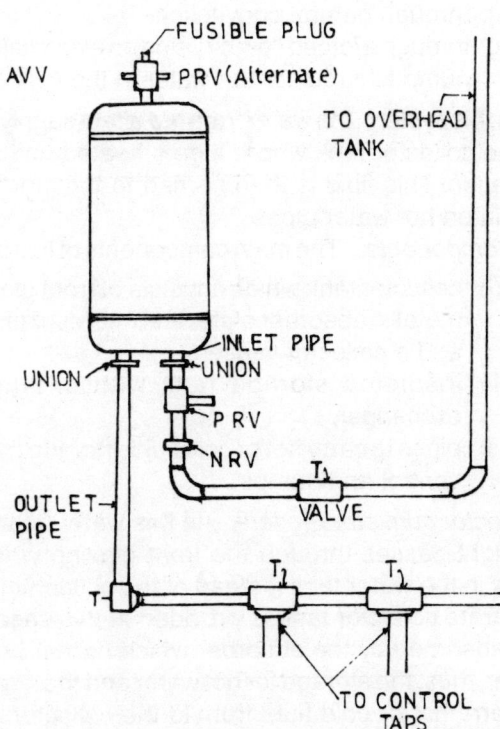

Fig. 12.1C *Sketch showing pressure type installation of geysers using pressure release valve.*

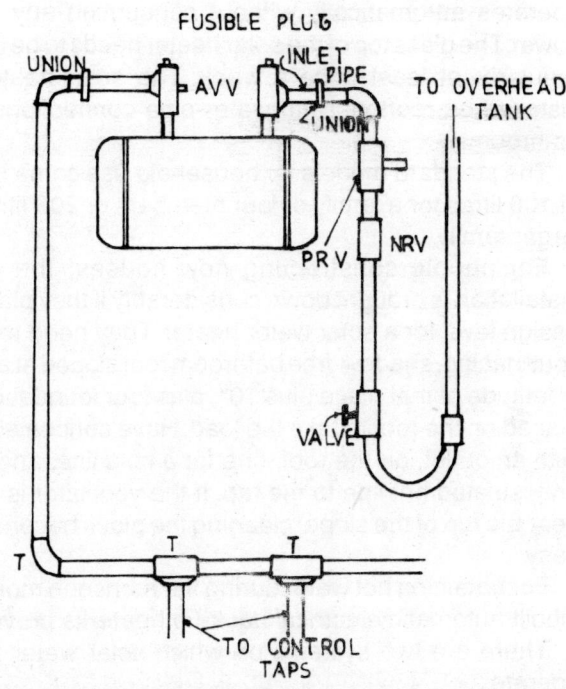

Fig. 12.1D *Pressure type installation using pressure release values using horizontal water heaters.*

There is another type of instant geyser—geyser cum shower (Figs 12.2A and B). The inlet of the geyser is connected to the water supply using a flexible pipe. The shower attachment is connected to the outlet. Flow control valve is adjusted by increasing or decreasing the water flow to set required water temperature.

The recent one is the instant shower, an advanced "geyser-cum shower". It works at the turn of the tap—turn the tap on, the electricity starts heating the water, and turn the tap off and the current supply is also turned off. The shower works even when the water pressure is low.

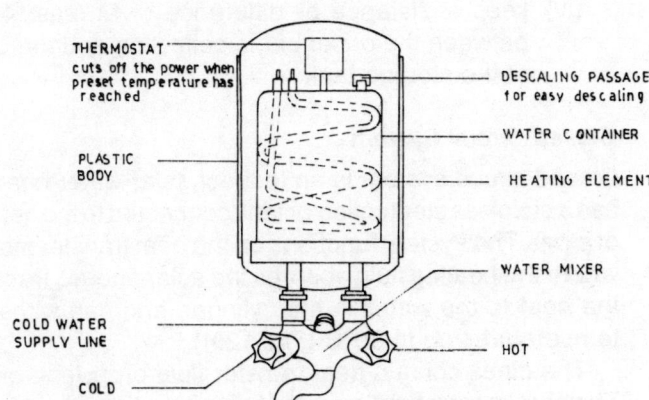

Fig. 12.2A *Sketch showing in built component of instant geyser.*

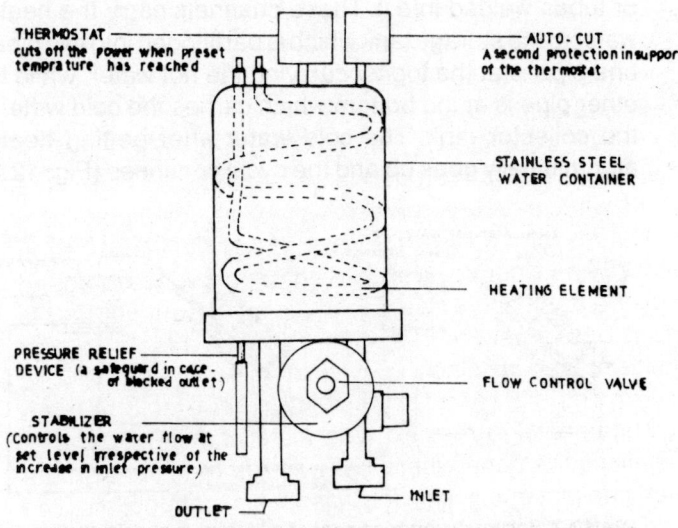

Fig. 12.2B *Sketch showing another type of instant geyser.*

Solar Water Heaters

As an energy conscious move, solar water heater can be an alternative to the electrically operated geysers for hot water supply in the bathrooms. The solar water heaters operate purely on solar energy—they absorb energy from the sun during the day and transfer it to water which is stored in a well-insulated stainless steel tank. The system operates automatically without consuming any electric power. The glass top of the solar heater needs to be cleaned regularly, at least once a week. The solar heaters are installed on rooftops with water-pipe connections to the bathrooms.

The standard models for household use come in range of 100 litres for a family of four members or 200 litres for a larger family.

For people constructing new houses, the cost of installation is brought down considerably if they plan at the design level for a solar water heater. They need to have a south facing, shadow-free bathroom roof sloped at an angle of latitude of that place plus 10°, plus four foundation bolts placed on the roof to take the load. Have concealed piping with an outlet, on the roof, one for a cold line, another for an insulated hot line to the tap. If the ventilator is located near the top of the slope, cleaning the glass becomes very easy.

For obtaining hot water during the monsoon months, an inbuilt automatic electrical back-up heater is provided.

There are two systems on which solar water heaters operate:

(i) open circuit system
(ii) closed circuit system.

Open Circuit System

The sunlight passes through the front glazing and gets absorbed by the absorber. The absorbed solar energy is converted into heat by the absorber plate. The heat is extracted from the absorber plate by the built-in channels or tubes welded into it. These channels carry the heated water to the storage tank which is partitioned into two pipes: one pipe is at the top for carrying the hot water, while the other pipe is at the bottom which carries the cold water to the collector tank. The cold water after getting heated automatically goes up and the cycle continues (Fig. 12.3).

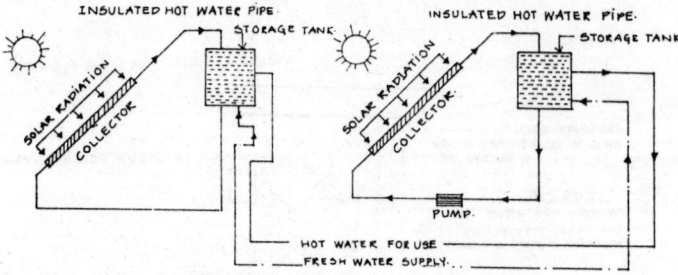

Fig. 12.3 *Schematic arrangement of a thermo-syphonic system. Solar water heater and arrangement of flow in solar water heater.*

Open circuit solar water heaters can be of two types:

(i) collector coupled to solar tank
(ii) collector cum storage system.

Collector coupled to solar tank   Heated water from the collector tank can be extracted in two ways:

(i) through natural convection
(ii) through a forced flow by using an electrically operated pump to circulate the water in the collector tank.

In this system, the water from the direct supply is pumped to the collector tank where it gets heated up by the solar radiation. This fluid is then carried to the storage tank in insulated hot water pipes.

Components   The main components of this system are:

(i) collector tank which consists of front glazing, black metallic absorber plate, thermal insulating material and a collector vessel,
(ii) insulated storage tank with or without heat exchanger,
(iii) pipes to carry hot or cold fluids from/to the collector,
(iv) control and pumps.

Collector cum storage tank   In this water heater system, sunlight passes through the front glazing which directly rests in the water tank instead of the collector tank, as no separate collector tank is provided in this case. Sun rays get absorbed by the absorber which further heats up the water, thus, the storage for hot water and the need for pipes to carry hot or cold fluid from/to the collector, as well as pumps and other controls is curtailed.

Components   The main components of this system are:

(i) front glazing
(ii) absorber sheet
(iii) insulated storage tank

Solar hot water systems can heat the water from ambient temperatures to about 85°C. To ensure good results, the collector should always:

(i) face within 15°C of true south
(ii) receive 6 hours of sunlight per day without shading
(iii) use GI pipes of suitable diameter to minimise pressure drops
(iv) keep a distance or difference of at least 40 cm between the outlet of the collection and the outlet of the storage tank.

Closed Circuit System

Closed circuit system is an indirect solar water heater. It has a stainless steel collector tank connected to the network of pipes. The system functions on the heat transfer method where the heating fluid absorbs the solar energy, transfers the heat to the water in the cylinder, and can withstand temperatures up to 80°C (Fig. 12.4).

The pipes contain heat transfer fluid propylene glycol. The heat transfer fluid is heated by a coated absorbed panel, then rises to indirectly heat the stainless steel storage tank.

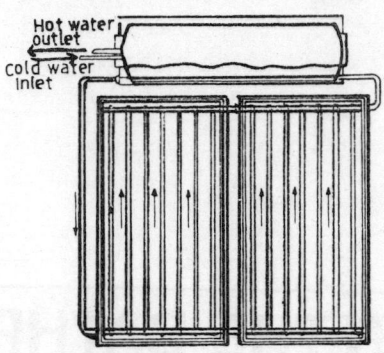

Fig. 12.4 *Diagram showing working of closed circuit, Solar water heater.*

The storage cylinder is sealed with a jacket (Fig. 12.5). This allows the closed circuit fluid to flow from the solar absorber panel around the outside of the storage cylinder and transfers its heat into the water stored in the cylinder. The cylinder is also properly insulated to keep the water hot for a longer time. The closed circuit system eliminates the risk of collector plates corroding or clogging from in side as they are not in direct contact with water.

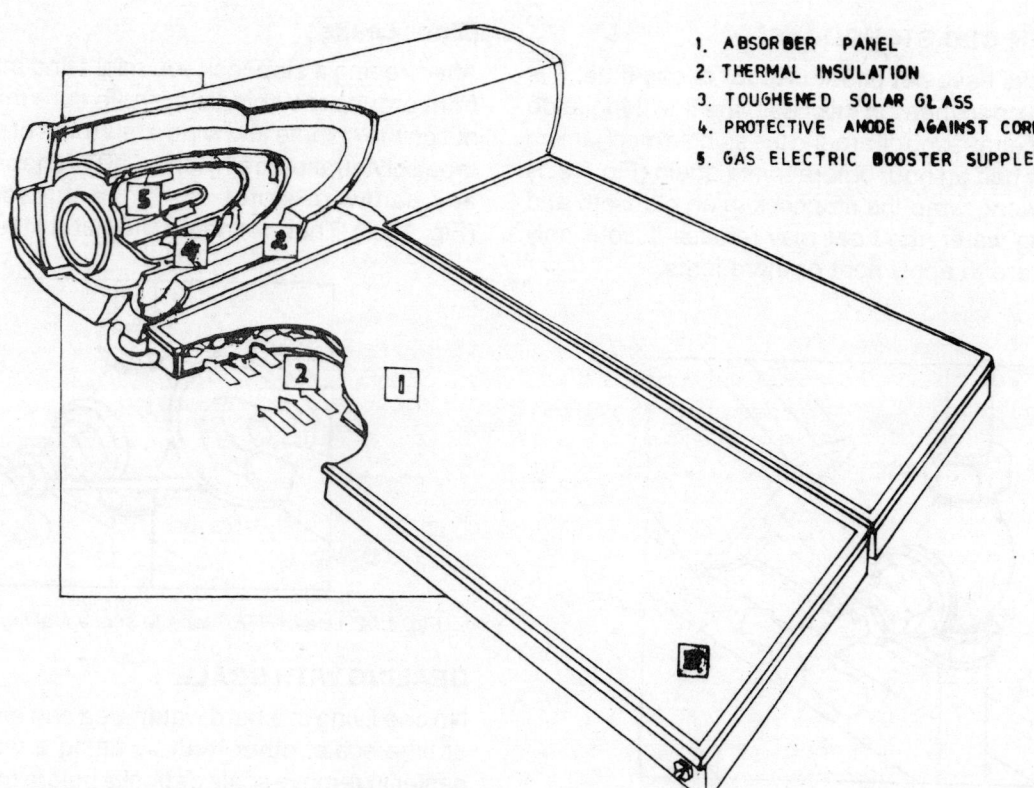

1. ABSORBER PANEL
2. THERMAL INSULATION
3. TOUGHENED SOLAR GLASS
4. PROTECTIVE ANODE AGAINST CORROSION
5. GAS ELECTRIC BOOSTER SUPPLEMENT

Fig. 12.5 *Detail of a closed circuit solar water heater.*

CHAPTER

13

PROBLEM AREAS IN THE BATHROOM

HOW TO OPEN OLD STOPCOCKS

When stopcocks have not been used for a long time, it is very difficult to open them. In this case, the first thing to do is squirt some penetrating oil around the spindle mechanism and leave it for half an hour before trying again (Fig. 13.1) If it does not work, wrap the stopcock in an old cloth and pour on boiling water, the heat may release it. Your only other options are to apply heat or more force.

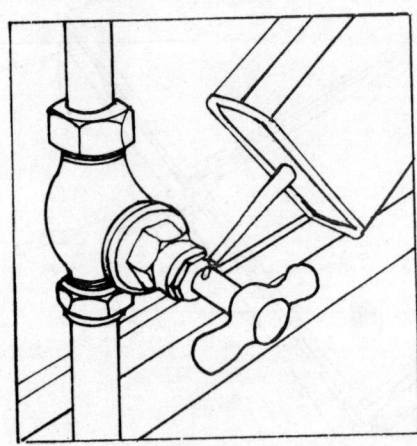

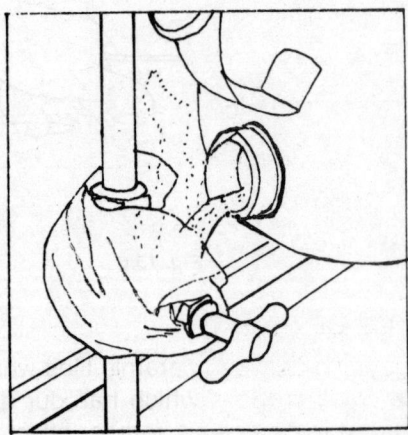

Fig. 13.1 Try freeing a seized stop cock with renetraling

Minor Leaks

After freeing a stopcock you might find that it leaks slightly from around the spindle area. To cure this, loosen the top nut on the spindle and wind a few turns of plumber's jointing tape polytetrafluoroethylene—(PTFE tape—available from any hardware store) round the threaded spindle body (Fig. 13.2). Then retighten the nut a little over hand tight.

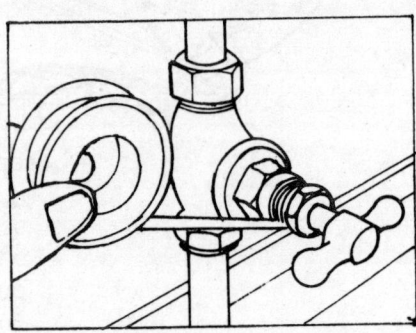

Fig. 13.2 Use of PTFE tape to seal a leaking stop cock spindle.

DEALING WITH SCALE

No one living in a hard water area can escape the probler of lime scale, other than by fitting a water softener. It i easier to remove scale deposits before they build up too fa There are two methods, the most effective being to app a special cleanser (Figs 13.3A and B). Some type however, are unsuitable for plastic appliances. Alternative use an old plastic card and a squeeze of lemon or vineg to chip off the scale (being plastic, the card will not scratc

Rust can also mark the surface of an appliance—of because something metallic has accidentally been lef contact with it. One way to remove scale is to chip it First, soak with lemon juice or vinegar, then use old pla card to chip away at the scale, working from the ed inwards.

CURING AIR LOCKS

Sometimes, when you turn the water back on, you he hissing and gurgling in the pipes and no water come of the taps. Usually, the pipes affected are those fed
he pressure is low

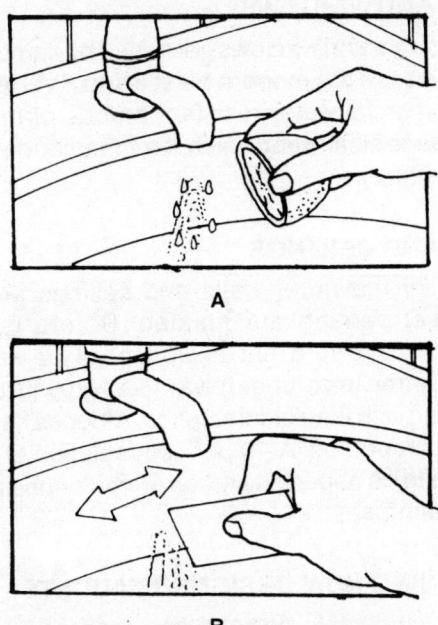

Figs 13.3A and B *Sketch showing how to remove scale, chip off the scale by use of old plastic card and squeeze of lemon or vinegar.*

Depending on the problem you may be able to force the air lock back up the pipe by taking a damp cloth and plunging vigorously against the spout of the affected tap, when the air reaches the tank, it will be able to escape (Fig. 13.4).

If this does not work, connect a hose between the affected tap and one that is supplied at mains pressure. Turn both taps on and leave them for about 30 seconds, then, turn the mains tap off and check to see if the air lock has cleared, sometimes it takes several goes (Fig. 13.5).

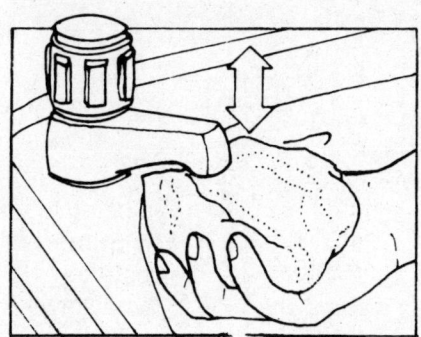

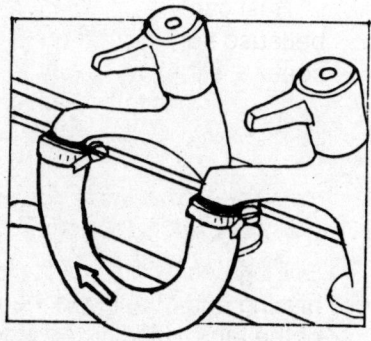

13.4 and 5 Sketch showing he ways of removal of air locking in taps.

Air locks can take ages to clear, so wherever possible, try to avoid them. While draining the water storage tanks, let the water back in as slowly as possible so that the air in the pipes has a chance to escape as they fill up. If you then get persistent air locks, trouble may be that the storage tank is not filling up as quickly as it is being drained, so the pipes are sucking in air instead of water.

RENEWING SEALANT

Like grout, lines of sealant along the tubs and basin often attract more dirt than the surfaces around them, and as a result start to look grubby. Some types of sealant mastic also crack or wrinkle over a period of time, harbouring still more dirt.

For a quick clean, use an old toothbrush dipped in household cleaner (or bathroom mould killer, if the sealant shows signs of black condensation mould).

A better solution is to renew the sealant. If the old sealant will not come off easily, clean and score the surface then apply a fresh layer.

To resurface old sealant, first clean it thoroughly, then score the surface. Fit the new sealant with a wide nozzle, then run it along the surface, working away from you (Fig. 13.6).

Use of Sealant sketch showing how to re-surface old sealant, first clean it thoroughly, then score the surface.

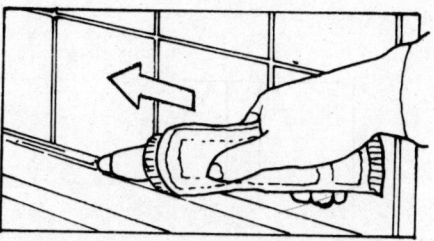

Fit the new sealant with a wide nozzle, then run it along the surface, working away from you.

Fig. 13.6

CLEANING UP GROUT

As the grouting between ceramic tiles wears with age, it develops small crevices which harbour dirt and germs. This in turn ruins the clean, regular effect of the tiles.

At one time, all you could do was to scrape out the grout and renew it. Now, grout colouring kits offer a much quicker way around the problem. They come in white, for freshening the original colour, and in a variety of shades it you fancy a change.

If the grout suffers from mould, use grout colouring, clean the tiles then brush on the colouring liquid following the grout lines (Fig. 13.7).

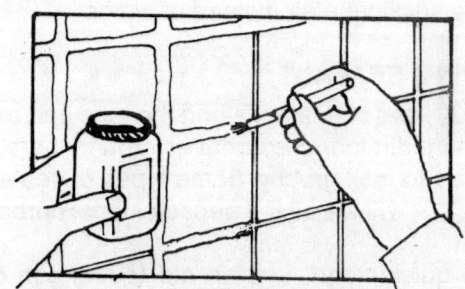

Fig. 13.7 Application of grout colouring clean the tiles then brush on the colouring liquid following the grout

Remove the excess liquid with a sponge and polish with a dry cloth.

CURING BLACK MOULDS

Lines of black mould around window frames and the edges of fittings are a common bathroom problem. The reason is usually condensation, which encourages mould growth and is almost impossible to cure completely. The problem is at its worst in winters, when windows are coldest. The best is to use mould killer. If you do not have one, clean with a weak solution of bleach (five parts water to one of bleach). An old toothbrush is the best tool to use, since it can reach into all the really awkward corners.

DOORS AND WINDOWS

Wooden doors and windows get ruined by dampness. Good alternatives are the modern plastic and PVC fittings or the new commercial water-resistant brands of marine ply. A polyurethane clear wood finish is recommended as a water repellent polish.

WALLS AND CEILINGS

Because of moisture, walls and ceilings are prone to dampness, peeling and flaking. Before going in for repainting, make sure that the surface of the walls is made completely free from fungal and mould growth, by brushing and treating with fungicidal solution. All loose flaky coatings must be scraped off. A 10 per cent solution of oxalic acid removes stains especially those of iron deposits from wall and floor surfaces.

MAINTAINING NEW DESIGNER FAUCETS

The most vulnerable parts of any sanitary fitting are the washers and the rings. These rubber or plastic parts are used to help prevent leakage between any two components of a fitting. In order to avoid wearing out of these parts, the handles of the fittings should not be tightened more than what is desirable. After using the faucet, ideally, the water droplets around the mouth should be dried. Use of acid to clean the surfaces is not recommended. Detergents which can react with the chemicals present in the plated or powder-coated surface in the presence of air, and water should not be used. Soap dust or chalk powder can be conveniently used to cleanse chrome plated taps. To clean powder-coated fittings, all one needs is just a smooth, dry cloth.

CHAPTER

14

WATERPROOFING TREATMENT

For the convenient plumbing arrangement under floor, the bathrooms have sunken floor slabs. This sunken portion contains the sloping soil pipe and floor trap. To avoid the dampness in the lower floors, this sunken portion should be properly treated with waterproofing material.

The main source of water leakage in the bathroom is the floor trap which dampens the RCC slab. However, if proper treatment to the sunken portion is not given, it is bound to fail if water keeps on leaking through the floor trap, affecting the aesthetics and hygiene of the lower floor.

Thus, the installation of the floor trap needs proper care. There are two methods by which the floor trap can be installed.

Using Multifloor Trap

Multifloor trap is designed to receive more than one connections of the waste pipe.

Laying Procedure of Multifloor Trap

Determine the correct location of the floor trap and set it on a firm base, located relative to the floor finish by pouring concrete on the slab.

The boss shoulders should be drilled out as required using a hole cutter, taking care not to damage surface of the shoulders.

Bedding can be carried out by pouring concrete around the floor trap, ensuring that the floor outlet is left clear of concrete.

The waste pipe can be inserted as required, and two surfaces can be sealed to make a permanent leakproof joint.

Adjusting the Height of the Floor trap in Regard to the Flooring

Remove floor trap top section, i.e. top tile and insert the raiser piece into body and seal it by waterproof compound. Replace removed floor trap top section on top of height raiser piece again, then seal it.

In places where multifloor trap cannot be used, 'there is another method by which the floor trap can be properly sealed. Determine the correct location of the floor trap and

place it on a firm base located relative to the floor finish by pouring concrete over the RCC slab.

Take an empty tin container which fits in the top opening of the floor trap and place the container so that its top is flushing with the floor level.

Drill holes in the container as required to take the waste pipe connections and insert the pipes into it and seal it properly with the waterproofing compound.

Pour rich concrete, with integral waterproofing compound, around the trap, with tin container acting as shuttering. The concrete should be well compacted, and there should be no honeycombing, so that the probability of water leakage through the concrete reduces to minimum.

Remove the container when the concrete has set and paint the internal surface of the concrete with a flexible waterproofing paint, like epoxy paint or acrylic paint. This paint shall seep into the spaces in between the aggregates. The waterproofing membrane so formed shall be flexible enough to sustain contraction and expansion of the concrete and will provide leakproof joint which will last long.

WATERPROOFING TREATMENT TO THE SUNKEN SLABS

Using Tapecrete

Application of 2 coats of "tapecrete"—acrylic polymer modified cementitious slurry coating on the sunken floor over the rendered surface and turned up and finished up to the floor level.

Tapecrete coating slurry
Mix proportion   100 kg cement
                     52 kg Tapecrete polymer

Material	1st coat on concrete kg/m^2	2 coates on concrete kg/m^2
Cement	0.488	0.730
P-151	0.253	0.379

Apply one coat of tapecrete—acrylic polymer modified cementitious brush topping over the tapecrete applied surface (Fig. 14.1).

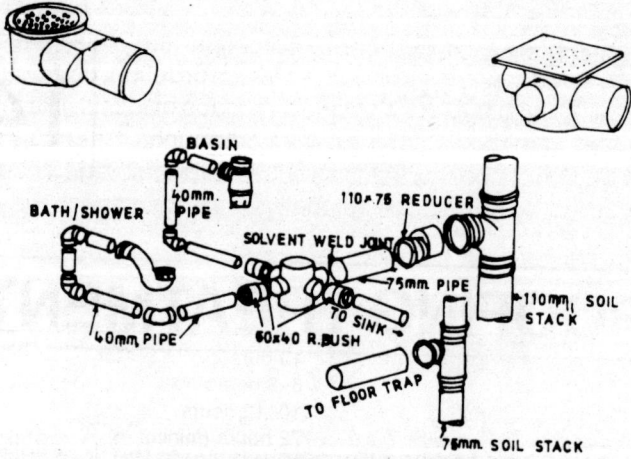

Fig. 14.1 *Typical layout of multifloor trap.*

Tapecrete Brush Topping

Material	Kg/m²	15 mm thick kg/m²
Cement	858	1.289
P-151 polymer	446	0.670
Fine silcia	855	1.289
	2162	

Filling the sunken portion with lime concrete/cement concrete upto the floor level (Fig. 14.2).

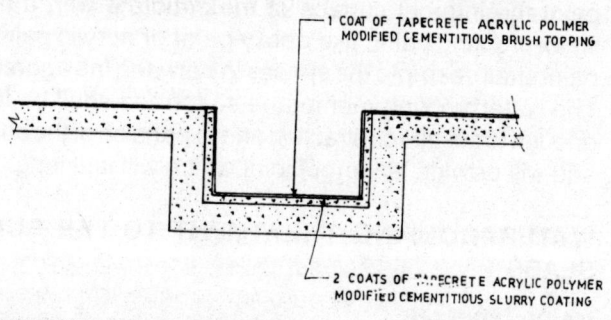

Fig. 14.2 *Details of water proofing in sunk in portion of bathrooms.*

Using Rubberised Waterproof Coating (RWPC)

RWPC is a single component—ready to use. Based on synthetic rubber, complex organic chemicals and rich minerals. RWPC is a brushable liquid, having paint consistency and is black in colour. RWPC can be applied to damp surfaces.

Surface Preparation

- Clean the surface of dust, loose particles, oils and grease
- Apply one coat of clear waterproofer as primer coat, let it dry for 4 to 6 hours

- Now all cracks including hairline cracks will be visible
- Fill the cracks of 1 to 10 mm width with joint filler compound with putty knife, spatula/finger. Avoid formation of cavities or bubbles. Allow to dry for 24 hours. Inspect the cracks if filled properly. If the shrinkage is observed, then top up once again with joint filler compound to the level of surface.

Method of Application

- Stir the contents thoroughly prior to use. Apply first coat of RWPC on prepared surface. Allow to dry for 8 hrs or overnight.
- Apply second coat of RWPC. If subsequent coats are required, they must be applied at intervals of 8 hours between each coat.

As RWPC is lighter in weight than water, it should not be applied if surface is water-logged or flooded. Do not apply on fresh concrete. Apply after 28 days of curing only.

Available in 1/2, 1, 4, 10, 20 litres under the tradename Chemistik RWPC.

Using Epoxy Slurry Mortar (ESM)

Epoxy slurry mortar is an instant waterproof coating which when applied stops leakages, dampness, seepage of water from internal or external masonary walls, RCC slabs, tanks, etc.

ESM is a three component product. This mortar can be applied on damp surfaces and is a brushable component. It is applied in bathroom, WC sunken slabs for permanent waterproofing of the area.

Surface Preparation

The surface preparation before application plays an important part in obtaining longer and durable performance from the coating. Clean the surface of loose particles and dust with the help of wire brush. Clean the surface oil, grease by using solvent such as trichloroethylene or by washing the area with detergent. Wash the area with water thoroughly to remove the contaminants. Allow to dry.

Chemical cleaning, if necessary, can be done by using hydrochloric acid commercial grade by diluting with water in ratio of 1:2 by volume. However, after treating surface, it is required to rinse/wash with water containing dilute ammonia solution (1 part ammonia diluted with 7 parts water).

Freshly laid concrete can be coated only after 30 days curing. Fill the cracks less than 10 mm with any crackseal compound and pot holes with floor resurfacer (A cementitious based grey powder mixed with binders and admixtures). All to cure overnight.

Method of Application

As the epoxy slurry mortar is three component consisting of:

(i) mortar resin
(ii) hardener
(iii) thickened water.

Add three components in the ratio 90:10:50 by weight. Stir three components vigorously for 5 minutes with iron rod or wooden stick. The contents will now turn into a brushable homogeneous mix.

Start applying on the prepared surface immediately. Use the material as fast as possible but within 2 hours as it will start curing and become hard in consistency. Allow to cure for 6 to 8 hours. Subsequent coats may be applied for longer durability.

Available in 1, 4, 10 kg packing as Chemistik epoxy slurry Mortar.

Using Coal Tar Epoxy

Coal tar epoxy is a solvent-based two component product evolved by blending of coal tar, epoxy resins with modified amine hardener and solvents.

Coal tar epoxy is highly elastic and can be applied on damp surface. Due to the extremely low water absorption property of tar, combined with exceptional adhesion property of epoxy resins, the coal tar epoxy coating is used for the sunken floor slab of bathrooms, also.

Surface Preparation

The surface should be clean, firm, free from loose particles, mortar droppings as well as free from oil, fats and grease.

Methods of Application

Coal tar epoxy is a cold setting 2 component product consisting of basic resin and hardener. Mix both component thoroughly in the ration 3:1 by volume. Avoid air entrapments during the mixing operations. Since the pot life of the mixture is limited, only those quantities should be mixed which can be used within 6 to 8 hours. The coating must not be applied at temperature below 10°C and in humidity more than 85 per cent as the setting process of applied material is a direct function of temperature. Intercoating of subsequent coats must be done while the previous coat is tacky or within 3 to 4 hours. The coating must be applied by a clean, dry nylon brush. Allow the coating to cure for at least 24 hours.

Precautions

Work with adequate ventilation and take care to avoid excessive contact of the compound with the skin or splashes entering into the eyes. Protect hands by wearing gloves or with correct barrier creams.

Available in 1, 4, 10, 20 litres packing as Chemistik coal tar epoxy.

Using Tar-Othane

Tar-othane is a two component, purified coal tar and polyurethane polymer-based waterproofing compound, which can be used for the sunken slabs of toilets. It is more durable than plain bituminous compounds. After curing it forms an impervious, tough, waterproof membrane. Though, black in colour, it does not melt like bitumen due to the sun heat.

Characteristics of Tar-othane

Application time (pot life)	: 1 hours
Drying time	: 6–8 hours
Recoatability time	: 10–12 hours
Complete curing time	: 72 hours (minimum)
Colour	: Black
Dry film thickness	: 40-45 microns/coat on concrete 20–25 microns/coat on metals
Coverage	: concrete surface 40-45 sq ft/lit/coat steel surface 70–75 sq ft/lit/coat
Paintability	: Aluminium paint can be applied.

Surface Preparation

The surface should be sound, dry, free of oil, grease and dust. Remove the loose particles by wire brushing. The new concrete/plaster should be cured with water for minimum 28 days.

Clean the dust by washing it with water. Allow the surface to dry. If the structure is too old and saturated with dirt, treatment with 15 to 20 per cent dilute hydrochloric acid is recommended with thorough wirebrush. Clean the dust by washing it with water, till neutralisation. Allow the surface to dry completely.

Application

- Seal the joints of pipes, walls to floors with polysulphide sealant. Allow it to cure for minimum 24 hours
- Apply one coat of epoxy coating as a primary coat
- Apply one or two coats of tar-othane. Sprinkle clean and dry silica sand on wet coat of tar-othane. Press the sand with trowel/hand. Allow it to cure for minimum 72 hours before doing brickbat coba. Use integral waterproofer in cement mortar of brickbat coba. Fix the tile with tile adhesive.

Precautions

- It is advisable to use hand gloves and goggles at the time of application.
- The room should be properly ventilated at the time of application.

Available in 1, 10, 20 kg packing manufactured by Choksey Chemicals Pvt. Ltd.

CHAPTER

15

INNOVATIVE IDEAS

Bathroom Doors

Bathroom doors disintegrate at the bottom after being in contact with water for a long time. To prevent this, the bath door should be made of a sound, water-resistant material or a good hardwood such as teak which can withstand the ravages of water to a certain extent. But sometimes even these start to wear out. A remedy for this is to cover the 6" on the exposed side, and preferably the bottom of the door as well, with a strip of aluminium or asbestos cement or any other material impervious to moisture (Fig. 15.1).

Folding Counters

For small bathrooms, to have an extra space for things like washing machine, a folding counter can be the answer. Figure 15.2 shows the folding counter, hinged parallel to the mirrored wall above the basin, leaving a space of few inches along the wall. This serves as a shelf for bottles, soap and other toiletries, which can stay put, even when the counter is lifted. The counter opens upwards without disturbing articles on the shelf behind.

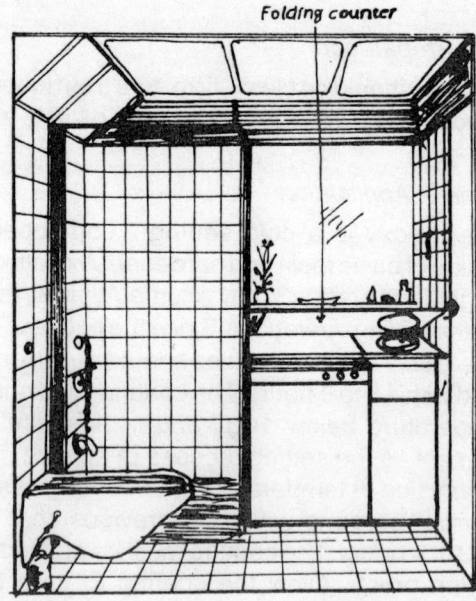

Fig. 15.2 Sketch shows use of folding counters.

Fig. 15.1 Sketch shows use of aluminium or asbestos strips as water ravages of doors.

Foot Scraper in the Flooring

Earlier, the floors of the bathrooms were kept rough to prevent it from becoming slippery from the oils used for the bath. The rough floor was also used as a foot scraper. Now, in the age of ceramic tiles with a smooth finish, you can have a foot scraper in your bathroom as well by embedding a piece of rough, unpolished Cuddapah stone in the bathing area. The size of the rough stone should be same as that of other tiles to fit into the flooring of the bathroom (Fig. 15.3).

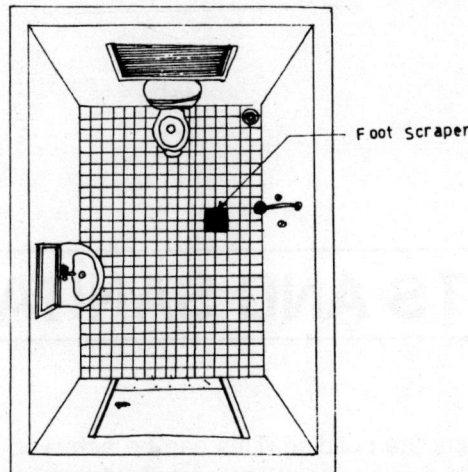

Fig. 15.3 *Top view of the bathroom showing location of foot scrapper.*

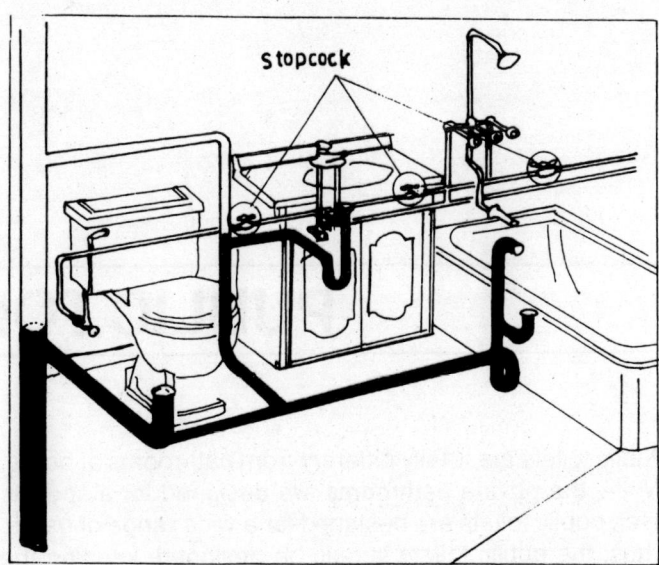

Fig. 15.4 *Sketch showing system of controlling water flow.*

Master Stopcock

The master stopcock is a device which controls the water supply to each bathroom, in addition to individual controls for the different sanitary fittings. The master stopcock makes the plumbing repairs very convenient, as when repairs are required, the water supply to the bathroom concerned can be easily turned off. In the absence of such a stopcock, it some times becomes necessary to shut off the main water supply from the overhead water tanks which, in a high rise can be quite difficult (Fig. 15.4).

PUBLIC TOILETS AND SANITATION

Public toilets are totally different from bathrooms at home. While, the private bathrooms are designed for a specific user, public toilets are designed for a wide range of users. Thus, the public toilets should be designed, keeping the common requirement in mind.

Privacy and hygiene are the most essential requirements of the public toilets as they are used by several user at the same time. At the designing stage, the location of the public toilet in a public building should be carefully decided so that they are easily seen and approachable. Separate toilets should be provided for males and females. The toilets should be well lit and ventilated, so that foul odours from the toilet should escape out as soon as possible and do not travel inside the building. This can be achieved by providing a buffer wall separating the toilet from other parts of the building, which would ensure privacy as well.

Design and facilities to be provided in the toilet depend on the type of the building, viz, commercial, industrial, hotels, hospitals, etc. and the number of users. Basic requirements of a public toilet are as follows.

Separate enclosure for WC counter with two or more washbasin, urinals for Gents toilets with partitions and squatting urinal enclosure for ladies toilets. There are other facilities that can be provided in the public toilet like electronic hand dryer for hygiene, automatic taps and automatic flushing cisterns (Figs 16.1A, B, C and 16.2A, B, C).

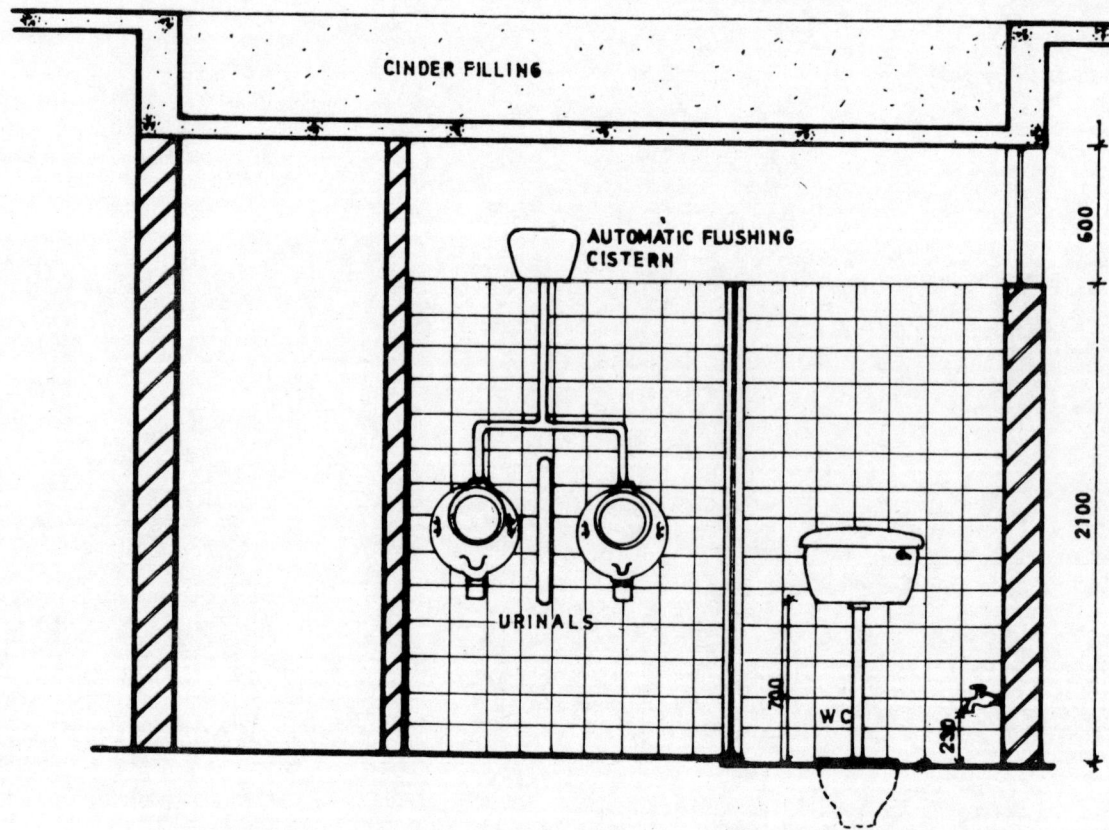

Figs 16.1A *Details of ladies and gents public toilets (small) showing plans and sections.*

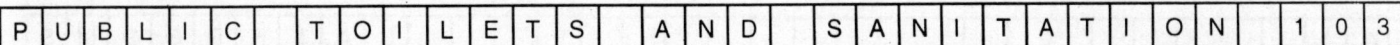

Figs 16.1B *Details of ladies and gents public toilets (small) showing plans and sections.*

CINDER FILLING

MIRROR

1000

HAND DRYER 300 LIQUID SOAP DISPENSER TOWEL RAIL

COUNTER BASINS

800-850

WC

410

390

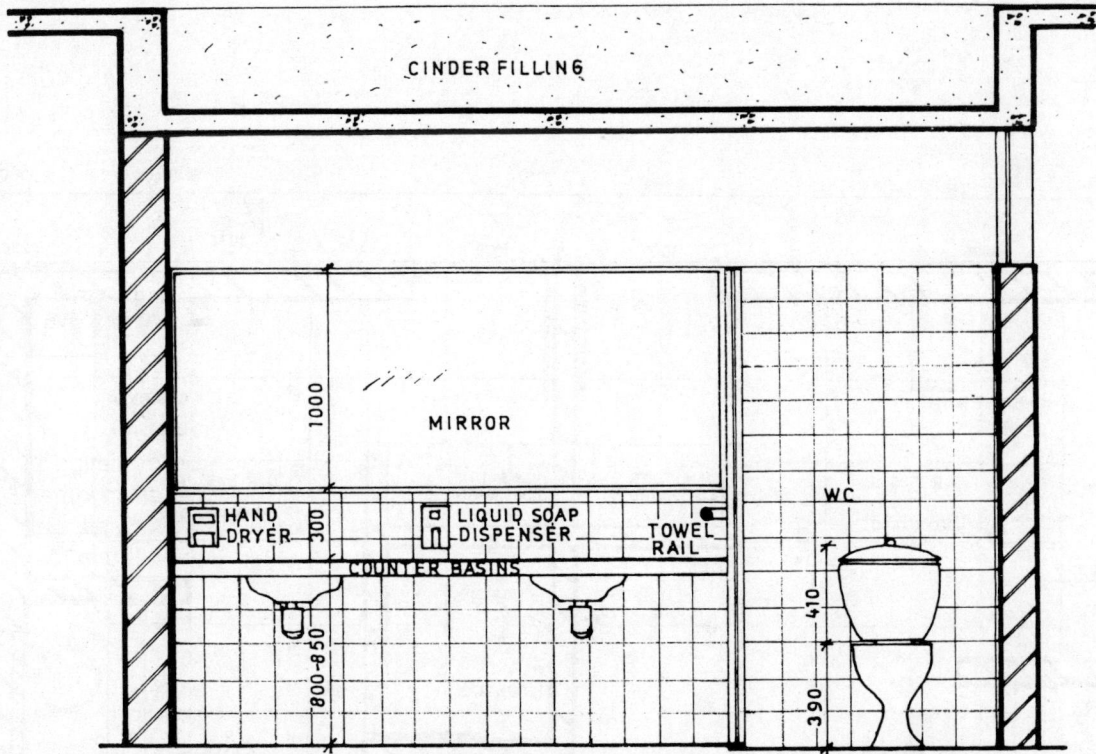

230   6715   230

230

900

D 1   D 1

300   900

450

1300   1100   1100   900   115

115

COUNTER BASIN

GENTS' TOILET
3000 X 3700

URINALS

D 2

LADIES' TOILET
3600 X 3700

COUNTER BASIN

800   750

20 670   1370

WC
1100 X 1370

450   115

680   1785   600

600   385   2000   400   1100

20 MM DROP
D 2

1500   D 2   1385   750   1200   D 2   1420
D 2

WC
1500 X 1200   WC
1385 X 1200   PIPE DUCT   WC
1200 X 1200   WC
1420 X 1200

1200   1200

R C C STONE   3700

230

1   PLAN   2

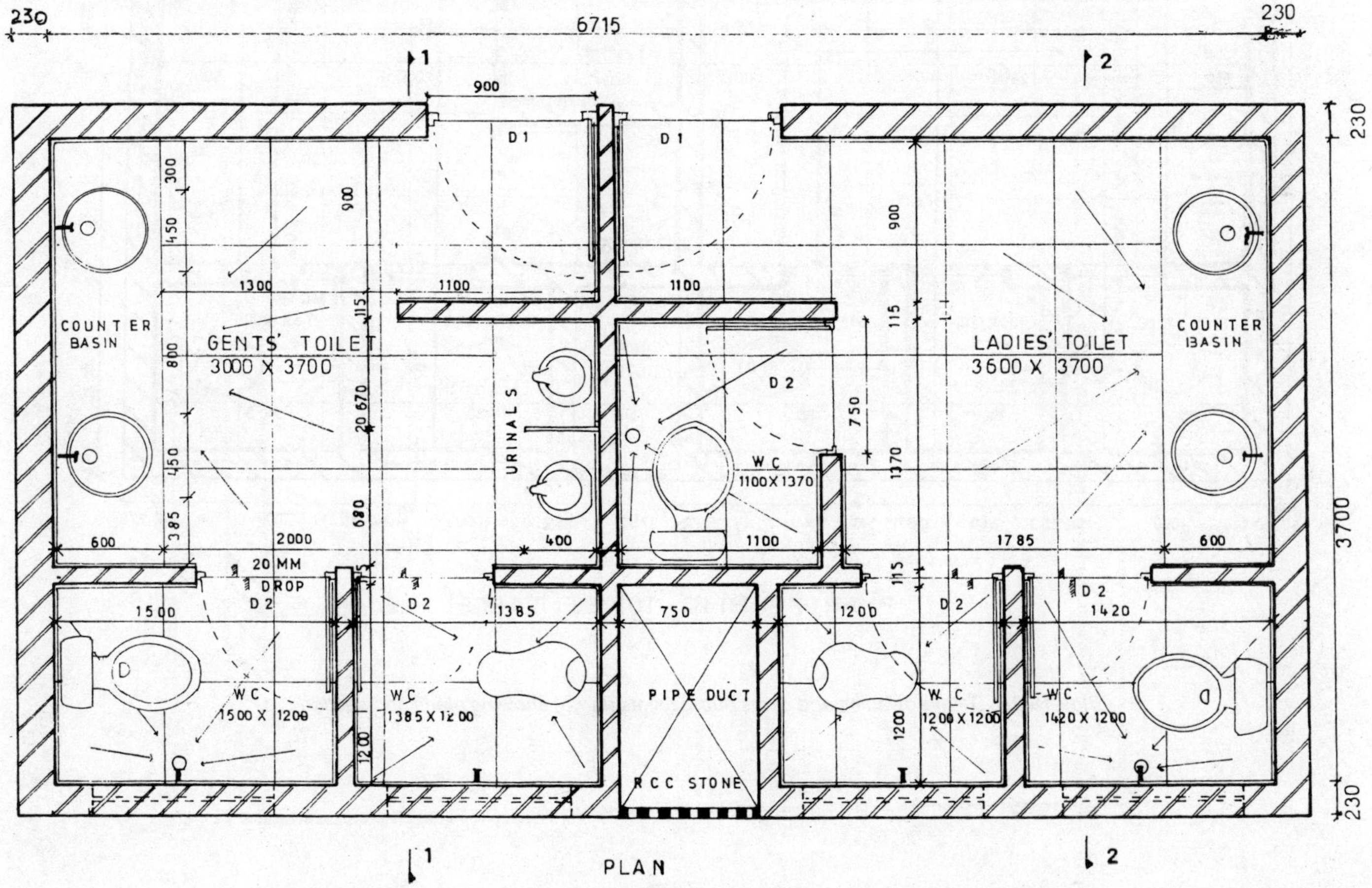

Figs 16.1C *Details of ladies and gents public toilets (small) showing plans and sections.*

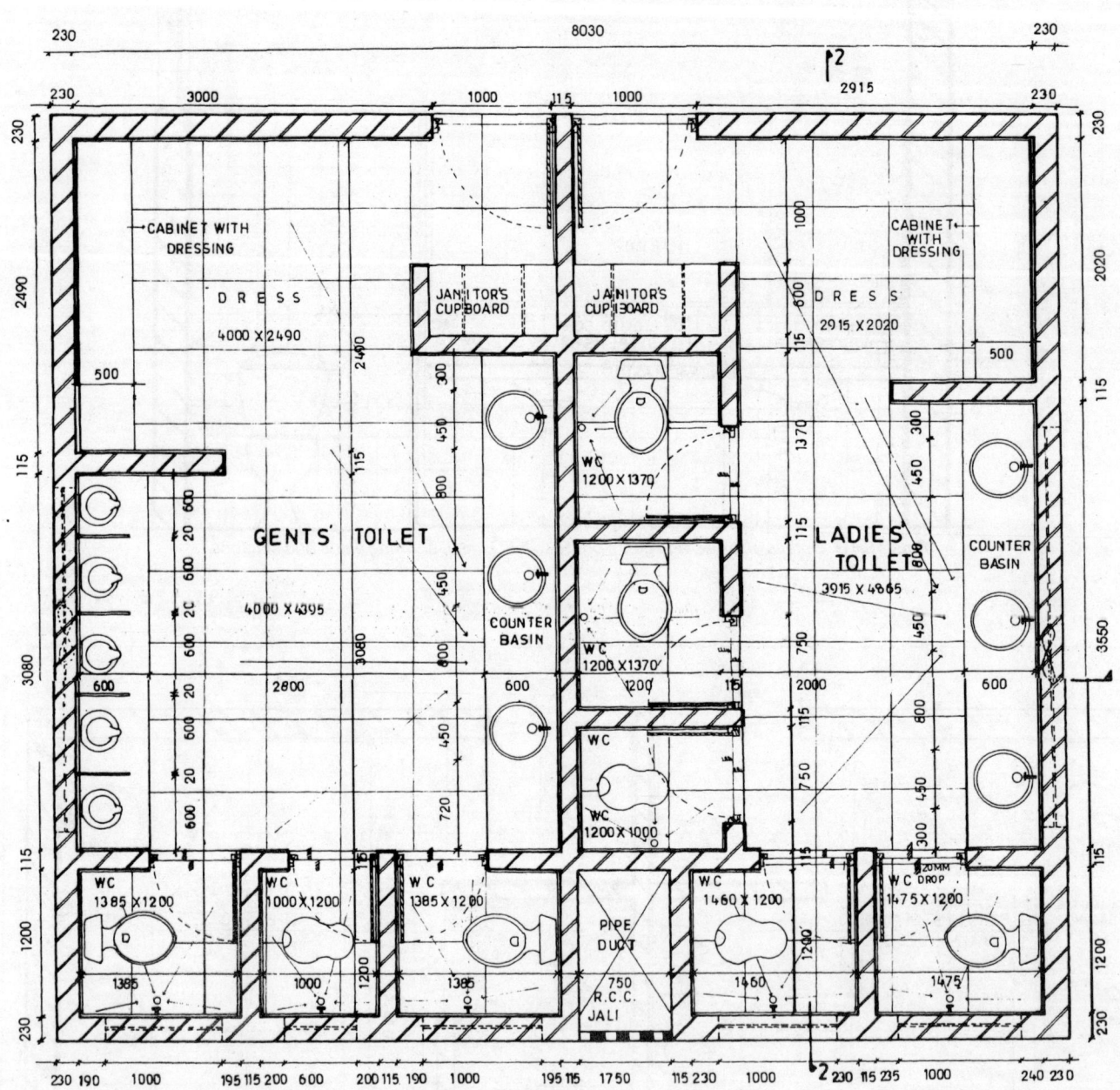

PLAN OF PUBLIC TOILET [LARGE]

Figs 16.2A *Details of ladies and gents public toilets (large) showing plans and sections.*

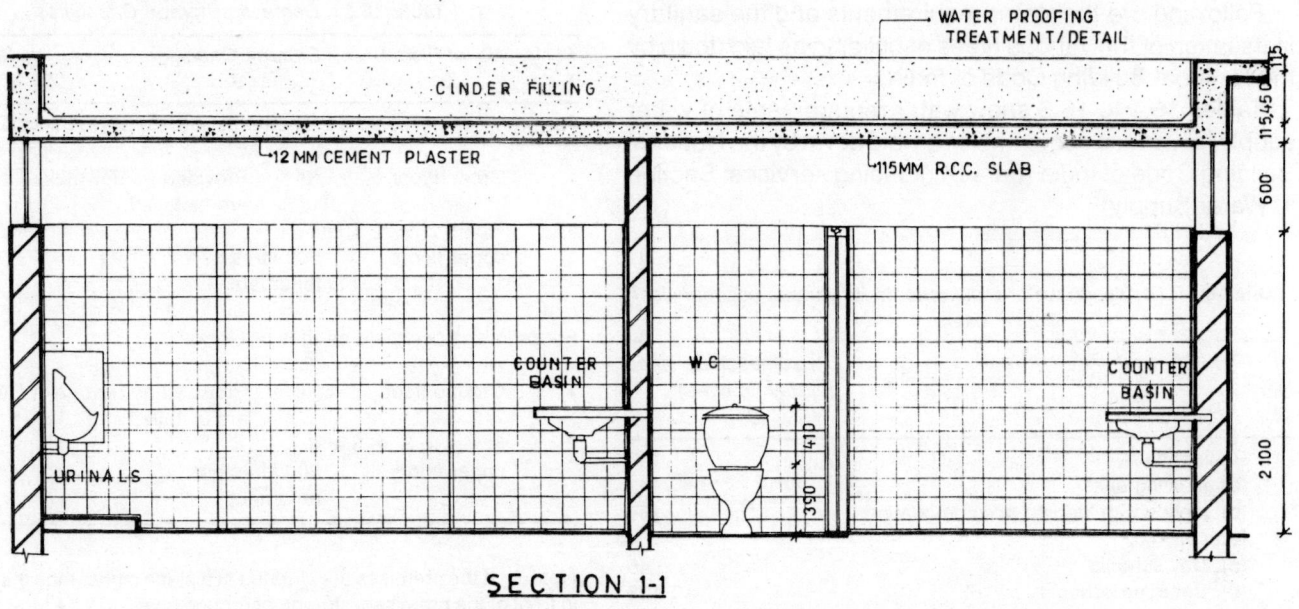

SECTION 1-1

Figs 16.2B Details of ladies and gents public toilets (large) showing plans and sections.

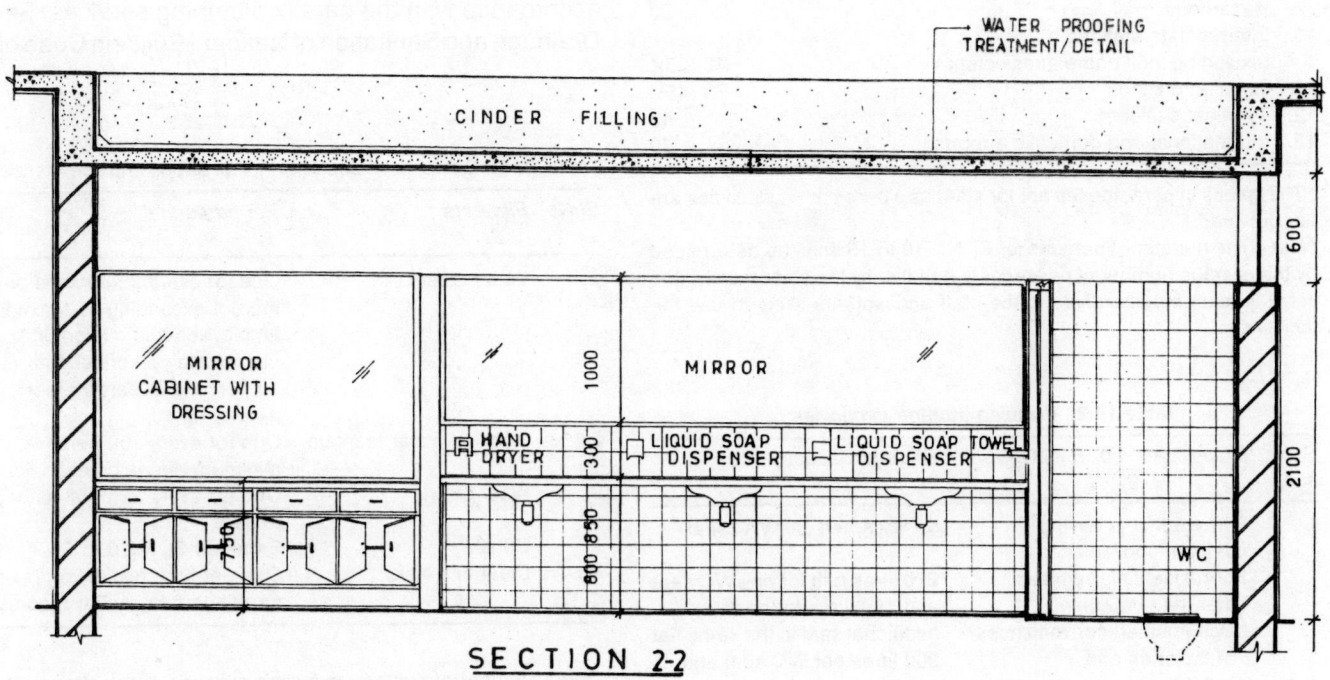

SECTION 2-2

Figs 16.2C Details of ladies and gents public toilets (large) showing plans and sections.

Following are the water requirements and the sanitary installation of the various types of building as laid down by the National Building Code of India.

Tables 16.1 to 16.3 show water requirements of water supply for various occupancies as laid down by the National Building Code of India (para IX plunding services: Section 1: Water Supply)

Table 16.1: *Per capita water requirements for various occupancies/users*

Sl. No. 1	Type of occupancy 2	Consumption per head per day (in litres) 3
1.	Residential:	135
	(a) in living units	
	(b) Hotels with lodging accommodation	180
2.	Educational:	
	(a) Day schools	45
	(b) Boarding schools	135
3.	Institutional (medical hospitals)	340
	(a) No. of beds not exceeding 100	350
	(b) No. of beds exceeding 100	450
	(c) Medical quarters and hotels	135
4.	Assembly-cinema theatres, auditoria, etc. (per seat of accommodation)	15
5.	Governmental or semipublic business	45
6.	Mercantile (commercial):	
	(a) Restaurants (per seat)	
	(b) Other business buildings	45
7.	Industrial:	
	(a) Factories where bathrooms are to be provided	45
	(b) Factories where no bathrooms are required to be provided	30
8.	Storage (including warehousing)	30
9.	Hazardous	30
10.	Intermediate stations (excluding mail and express stops)	45 (25)*
11.	Junction stations	70 (45)*
12.	Terminal stations	45
13.	International and domestic airports	70

* The values in parentheses are for stations where bathing facilities are not provided.

Note The number of persons for Sl. No. 10 to 13 shall be determined by the average number of passengers handled by the station daily, due consideration may be given to the staff and workers likely to use the facilities.

Table 16.2: *Flushing storage capacities*

Sl.No. 1	Classification of Buildings 2	Storage Capacity 3
1.	For tenements having common convenience	90 litres nett. per WC seat
2.	For residential premises other than tenement having comon conveniences	270 litres nett for one WC seat and 180 litres for each additional seat in the same flat
3.	For factories and workshops	900 litres per WC seat and 180 litres per urinal seat
4.	For cinemas, public assembly halls, etc.	900 litres per WC seat and 350 litres per urinal seat.

Table 16.3: *Domestic Storage Capacities*

Sl No 1	No. of Floors 2	Storage Capacity 3	Remarks 4
For premises occupied as tenements with common conveniences:			
1.	Groundfloor	Nil	Provided no downtake fittings are installed
2.	Floors 2, 3, 4, 5 and upper floors	500 litres per tenement	—
For premises occupied as flats or blocks:			
1.	Ground floor	Nil	Provided no downtake fittings are installed
2.	Floors 2, 3, 4, 5 and upper floors	500 litres per tenement	

Note 1 If the premises are situated at a place higher than the road level in front of the premises, storage at ground level shall be provided on the same lines as on floors.

Note 2 The above storage may be permitted to be installed provided that the total domestic storage calculated on above basis is not less than the storage calculated on the number of downtake fittings according to the scales given below:

Downtake taps	70 litres each
Showers	135 litres each
Bathtubs	200 litres each

Tables 16.4 to 16.15 show the requirement for sanitary fittings and installations for various occupancies in accordance with the para ix plumbing services: Sector-2: Drainage and Sanitation of National Building Code of India.

Table 16.4: *Sanitation Requirements for shops and commercial offices*

Sl.No. 1	Fitments 2	For personnel 3
1.	Water closet	One for every 25 persons or part thereof exceeding 15 (including employees and customers). For female personnel 1per every 15 persons or part thereof exceeding 10
2.	Drinking water fountain	One for every 100 persons with a minimum of one on each floor
3.	Washbasin	One for every 25 persons or part thereof
	Urinals	Same as Sl. No. 3 to Table 15.
4.	Cleaner's sink	One per floor minimum preferably in or adjacent to sanitary rooms.

Note Number of customers for the purpose of the above calculation shall be the average number of persons in the premises for a time interval of one hour during the peak period. For male-female calculation a ratio 1:1 may be assumed.

Table 16.5: *Sanitation requirements for large stations and airports*

Sl.No. 1	Place 2	WC for Males 3	WC for Females 4	Urinals for Males only 5
1.	Junction stations, intermediate stations and bus stations	3 for first 1000 persons and 1 for every subsequent 100 persons or part thereof	4 for first 1000 persons and 1 for every addl. 1000 persons or part thereof	4 for every 1000 persons and 1 for every additional 100 persons or part thereof
2.	Terminal stations, and bus terminals	4 for first 1000 persons and 1 for every subsequent 1000 persons or part thereof	5 for first 1000 persons and 1 for every subsequent 1000 persons or part thereof	6 for first 1000 persons and 1 for every additional 1000 persons or part thereof
3.	Domestic airports			
	Min.	2*	4*	2*
	for 200 persons	5	8	6
	for 400 persons	9	15	12
	for 600 persons	12	20	16
	for 800 persons	16	26	20
	for 1000 persons	18	29	22
4.	International airports			
	for 200 persons	6	10	8
	for 600 persons	12	20	16
	for 1000 persons	18	29	22

Note Provision for washbasins, baths including shower stalls, shall be in accordance with Part IX Section 2 Drainage and Sanitation of National Building Code of India.
* At least one Indian style water closet shall be provided in each toilet. Assume 60 males to 40 females in any areas

Table 16.6: *Sanitary Requirements for Hotels*

Sl. No. 1	Fitments 2	For Residential public and staff 3	For Public Rooms		For Nonresidential staff	
			For males 4	For females 5	For Males 6	For Females 7
1.	Water closet	One per 8 persons excluding occupants of the room with attached water closet minimum of 2 if both sexes are lodged	One per 100 persons up to 400 persons: for over 400 add at the rate of one per 250 persons or part thereof	2 for 100 persons up to 200 persons; over 200 add at the rate of one per 100 persons or part thereof	1 for 1–5 persons 2 for 16–35 persons 3 for 36–65 persons 4 for 66-100 persons	1 for 1–12 persons 2 for 13–25 persons 3 for 26–40 persons 4 for 41–57 persons
2.	Ablution taps	One in each water closet	One in each water closet	One in each water closet	One in each water closet	One in each water closet
		1 water tap with drainage arrangements shall be provided for every 50 persons or part thereof in the vicinity of water closet and urinals				
3.	Urinals	—	One for 50 persons or part thereof	—	Nil up to 6 persons 1 for 7–20 persons 2 for 21–45 persons 3 for 46–70 persons 4 for 71–100 persons	
4.	Washbasins	One per 10 persons omitting the wash-basins installed in the room suits	One per water closet and urinal provided	One per water closet provided	1 for 1–15 persons 2 for 16–35 persons 3 for 36–65 persons 4 for 66–100 persons	1 for 1–12 persons 2 for 13–15 persons 3 for 26–40 persons 4 for 41–57 persons 5 for 58–77 persons
5.	Baths	One per 100 persons omitting occupants of the room with bath in suits,	—	—	—	—
6.	Sop sinks	One per 30 bedroom (one per floor min)	—	—	—	—
7.	Kitchen sinks	One in each kitchen	One in each kitchen	One in each kitchen	One in each kitchen	One in each kitchen

Note It may be assumed that the two-third of the number of males and one-third females

Table 16.7: *Sanitation Requirements for Educational Occupancy*

Sl. No. 1	Fitments 2	Nursery schools 3	Boarding institution		Other educational institutions	
			For Boys 4	For girls 5	For boys 6	For girls 7
1.	Water closet	One per 15 pupils and part thereof	One every 8 pupils or part thereof	One very 6 pupils or part thereof	One/40 pupils or part thereof	One/25 pupils or part thereof
2.	Ablution taps	One in each water closet	One in each water closet	One in each water closet	One in each water closet	One in each water closet
		One water tap with draining arrangements shall be provided for every 50 persons or part thereof, in the vicinity of water closets and urinals				
3.	Urinals	—	One per every 25 pupils or part thereof	—	One per every 20 pupils or part thereof	
4.	Washbasins	One per 15 pupils or part thereof	One for every 8 pupils or part thereof	One for every 6 pupils or part thereof	One per 40 pupils or part thereof	One per 40 pupils or part thereof
5.	Baths	One bath sink per 40 pupils	One for every 8 pupils of part thereof	One for every 6 pupils or part thereof	—	—
6.	Drinking water fountains	One for every 50 pupils or part thereof	One for every 50 pupils or part thereof	One for every 50 pupils or part thereof	One for every 50 pupils or part thereof	One for every 50 pupils or part thereof
7.	Cleaner's sink	—	One per floor minimum	One per floor minimum	One per floor minimum	One per floor minimum

Note For teaching staff, the schedule of fitments to be provided shall be the same as in the case of office buildings

Table 16.8: *Sanitation requirements for institutional (medical) occupancy-hospitals*

Sl. No. 1	Fitments 2	Hospitals with Indoor patient wards for males and females 3	Hospitals with outdoor patient wards		Administrative Buildings	
			For males 4	For females 5	For male Personnel 6	For female Personnel 7
1.	Water closet	One for every 6 bed	One for every 100 persons or part thereof	Two for every 100 persons of part thereof	One for every 25 persons or part thereof	One for every 15 per persons or part thereof
2.	Ablution tap	One in each water	One in each water closet	One in each water closet	One in each water closet	One in each water closet
		One water tap with draining arrangements shall be provided for every 50 persons or part thereof, in the vicinity of water closet and urinals				
3.	Washbasins	2 upto 30 beds; add one for every additional 30 beds; or part thereof	One for every 100 persons of part thereof	One for every 100 persons or part thereof	One for every 25 persons or part thereof	One for every 25 persons or part thereof
4.	Baths with shower	One bath with shower for every 8 beds or part thereof	—	—	One on each floor	One on each floor
5.	Bedpan washing sinks	One for each ward	—	—	—	—
6.	Cleaner's sinks	One for each ward	One per floor minimum	One per floor minimum	One per floor minimum	One per floor minimum
7.	Kitchen sinks and dish washers (where kitchen is provided)	One for each ward	—	—	—	—
8.	Urinals	—	One for every 50 persons or part thereof	—	Nil up to 6 persons One for 7–20 persons 2 for 21–45 persons 3 for 46–70 persons 4 for 71–100 persons from 101–200 persons add at the rate of 3%, for over 200 persons add at the rate of 2.5%	

Table 16.9: *Sanitation requirements for institutional medical occupancy (staff quarters and hostels)*

Sl. No.	Fitments	Doctor's dormitories		Nurse's Hostel
		For male staff	For female staff	
1	2	3	4	5
1.	Water closets	One for 4 persons	One for 4 persons	One for 4 persons or part thereof
2.	Ablution taps	One in each water closet	One in each water closet	One in each water closet
3.	Washbasins	One for every 8 persons or part thereof	One for every 8 persons or part thereof	One for every 8 persons or part thereof
4.	Bath (with shower)	One for 4 persons or part thereof	One for 4 persons or part thereof	One for 4 to 6 persons or part thereof
5.	Cleaner's sinks	One per floor minimum	One per floor minimum	One per floor minimum

Table 16.10: *Sanitation requirements for governmental and public business occupancies and offices*

Sl. No.	Fitments	For male personnel	For Female personnel
1	2	3	4
1.	Water closets	One for every 25 persons or part thereof	One for every 15 persons or part thereof
2.	Ablution taps	One in each water closet	One in each water closet
		One water tap with draining arrangements shall be provided for every 50 persons or part thereof, in the vicinity of water closet and urinals	
3.	Urinals	Nil up to 6 person one for 7–20 persons 2 for 21–45 persons 3 for 46–70 persons 4 for 71–100 persons From 101–200 persons add at the rate of 3% For over 200 persons add at the rate of 2.5%	
4.	Washbasins	One for every 25 persons or part thereof	
5.	Drinking water fountains	One for every 100 persons, with a minimum of one on each floor	
6.	Baths	Preferably one on each floor.	
7.	Cleaner's sinks	One per floor minimum preferably in or adjacent to sanitary rooms	

Table 16.11: *Sanitation requirements for residences*

Sl. No.	Fitments	Dwellings with individual conveniences	Dwellings without individual conveniences
1	2	3	4
1.	Bathroom	1 provided with water tap	1 for every two tenements
2.	Water closets	1	1 for every two tenements
3.	Sink (or nahani) in the floor)	1	—
4.	Water tap	1	1 with draining arrangement in each tenement 1 in common bathrooms and common water closets

Note Where only one water closet is provided in a dwelling, the bath and water closet shall be separately accommodated

Table 16.12: *Sanitation requirements for assembly occupancy buildings (Cinema, theatres, auditoria, etc.)*

Sl. No.	Fitments	For public		For staff	
		Male	Female	Male	Female
1	2	3	4	5	6
1.	Water closets	1 per 100 person up to 400 persons. For over 400 persons, add at the rate of 1 per 250 persons or part thereof	2 per 100 persons up to 200 persons For over 200 persons, add at the rate of 1 per 100 persons or part thereof	1 for 1–15 persons 2 for 16–35 persons	1 for 1–12 persons 2 for 13–25 persons
2.	Ablution taps	1 in each water closet	1 in each water closet	1 in each water closet	1 in each water closet
		One water tap with draining arrangements shall be provided for every 50 persons or part thereof in the vicinity of water closets and urinals.			
3.	Urinals	1 for 50 persons or part thereof —		Nil up to 6 persons 1 for 7–20 persons 2 for 21–45 persons	
4.	Washbasins	1 for every 200 persons or part thereof	1 for every 200 persons or part thereof	1 for 1–15 persons 2 for 16–35 persons	1 for 1–12 persons 2 for 13–25 persons

Note It may be assumed that two-third of the number are males and one-third females

Table 16.13: *Sanitation requirements for assembly buildings (art galleries, libraries and museums)*

Sl. No.	Fitments	For public		For staff	
		Male	Female	Male	Female
1	2	3	4	5	6
1.	Water closets	1 per 200 persons upto 400 persons. For over 200 persons, add at the rate of 1 per 250 persons or part thereof.	1 per 100 persons upto 200 persons. For over 200 persons. add at the rate of 1 per 150 persons or part thereof	1 for 1–15 persons 2 for 16–35 persons.	1 for 1–12 persons 2 for 13–25 persons
2.	Ablution taps	1 in each water closet	1 in each water closet	1 in each water closet	1 in each water closet
		One water tap with draining arrangements shall be provided for every 50 closet, persons or part thereof in the vicinity of waters closets and urinals,			
3.	Urinals	1 for 50 persons		Nil up to 6 persons 1 for 7–20 persons 2 for 21–45 persons	
4.	Washbasins	1 for every 200 persons or part thereof. For over 400 persons, add at the rate of 1 per 250 persons or part thereof	1 for every 200 persons or part thereof. For over 200 persons, add at the rate of 1 per 150 persons or part thereof.	1 for 1–15 persons 2 for 16–35 persons	1 for 1–12 persons 2 for 13–25 persons
5.	Cleaner's sink	.... 1 per floor, minimum, ...			

Note If may be assumed that two-third of the number are males and one-third females

Table 16.14: *Sanitation requirements for restaurants*

Sl. No.	Fitments	For public		For staff	
		Male	Female	Male	Female
1	2	3	4	5	6
1.	Water closets	One for 50 seats up to 200 seats. For over 200 seats, add at the rate of one per 100 seats or part thereof	One for 50 seats up to 200 seats. For over 200 seats, add at the rate of one per 100 seats or part thereof	One for 1–15 persons 2 for 16–35 persons 3 for 35–65 persons 4 for 66–100 persons	1 for 1–12 persons 2 for 13–25 persons 3 for 26–40 persons 4 for 41–57 persons 5 for 58–77 persons 6 for 78–100 person
2.	Ablution taps	One in each water closet	One in each water closet	One in each water closet	One in each water closet
		One water tap with draining arrangement shall be provided for every 50 persons or part thereof in the vicinity of water closets and urinals			
3.	Urinals	One per 50 seats	—	Nil upto 6 persons 1 for 7–20 persons 2 for 21–45 persons 3 for 46–70 persons 4 for 71–100 persons	
4.	Washbasins	.... One for every water closet provided			
5.	Kitchen sinks and dish washer	.... One in each kitchen			
6.	Slop or service sink	.... One in the restaurant			

Note If may please be assumed that two-third of the number are males and one-third females

Table 16.15: *Sanitation requirements for factories*

Sl. No. 1	Fitments 2	For male personnel 3	For female personnel 4
1.	Water closets	1 for 1–15 persons 2 for 16–35 persons 3 for 36–65 persons From 101 to 200 persons, add at the rate of 3% From over 200 persons, add at the rate of 2.5%	1 for 1–12 persons 2 for 13–25 persons 3 for 26–40 persons 6 for 78–100 persons From 101 to 200 persons, add at the rate of 5% From over 200 persons, add at the rate of 4%
2.	Ablution taps	1 in each water closet. One water tap with draining arrangement shall be provided for every 50 persons or part thereof in the vicinity of water closet and in urinals	1 in each water closet
3.	Urinals	Nil up to 6 persons 1 for 7–20 persons 2 for 21–45 persons 3 for 46–70 persons 4 for 71–100 persons From 101–200 persons, add at the rate of 3% for over 200 persons, and at the rate of 2.5%	
4.	Washing taps with draining arrangements	1 for every 25 persons or part thereof	1 for every 25 persons or part thereof
5.	Drinking water fountains	1 for every 100 persons with a minimum of one on each floor	
6.	Baths preferably showers	As required for particular trades or occupations	

Note For many trades of a dirty or dangerous character, more extensive provisions are required

RURAL SANITATION

The main factors to be emphasised in any rural sanitation programme are as follows:

1. Diminishing the risk to health caused by intestinal and helminthic infections
2. The social dimension of liberating scavengers from a demeaning job
3. Economic factors to be considered are:
   a. Saving of energy and recycling it to get fertiliser and gas
   b. Decreasing consumption through more efficient use of available energy resources,
   c. The cost of intallation and maintenance and the possibility of generating employment through the projects, are also important issues for a successful programme.

Sanitary Toilets

No ideal fool-proof solution has yet been found for low-cost sanitary disposal of human waste. Each system has some disadvantages/drawbacks either in the short-term or long-term sense. But on the basis of studies and research conducted so far, certain guidelines are available. Efforts to introduce intermediate technology are aimed at ameliorating conditions of rural life in keeping with modern civilisation—not perfecting them at one go.

There are two main disposal systems available for rural areas in terms of technical, economic and social feasibility. They are:

(i) leaching pits
(ii) septic tanks.

To avoid the risk of pollution of ground water, it is advised by NEERI (National Environmental Engineering Research Institute) that a leaching pit should be sited not less than six metres downstream from a well or water source and that the subsoil water should be at least one metre below the bottom of the pit. These distances can very according to the type of soil. Pollution travels less in clayey soils and more in sandy soils. With longer use, the pores in the pit get filled up, and the distance of pollution travel decreases.

Maturing Period for Sludge

With regard to all the leaching pit designs, the period allowed for completion of the decomposition process or "maturing"—before the pit is opened and the sludge handled is—a one-year maturing period, the minimum acceptable being six months.

Various designs of sanitary latrines as put forward by various agencies involved in rural sanitation are as follows.

"Sopa Sandas": A Ventilated Improved Double Pit Latrine (Fig. 16.3)

"Sopal" type is a ventilated double-pit privy with a tin flap instead of a water-seal trap. The components of the privy are:

(i) RCC or stone slab with cement pan
(ii) steep sloping pipe with a tin flap at the upper end
(iii) rectangular displaced pit with partial honeycomb brick-lining
(iv) t-pipe to connect both pits
(v) a vent pipe to carry away odours from the pit. The flap gets worn out after several years, but it can easily be replaced at minimal cost.

If pits are maintained and handled according to specifications, the matured sludge is totally harmless and inoffensive. It can be extracted by the house-holder himself and yields valuable manure. This type of latrine is used in Maharashtra.

VIP (Ventilated Improved Pit) Latrine (Fig. 16.4)

Ventilated improved single-pit type based on a design developed in Rhodesia. Aerobic action takes place in the pit through a cycle provided by air suction through the toilet seat and up the vent pipe, which is warmed by the sun and therefore draws up the air. Rain water should not be allowed to enter. The toilet seat should be kept in the dark so as not to attract flies. The vent pipe should be at least 6" diameter.

This type of latrine has been designed for use in Punjab. Since no flushing water is required, contents of the pit are relatively dry and therefore pollution travel is less. This is an important factor in Punjab, where most households have their own handpumps and therefore keeping greater distance between the toilet and well can become a problem.

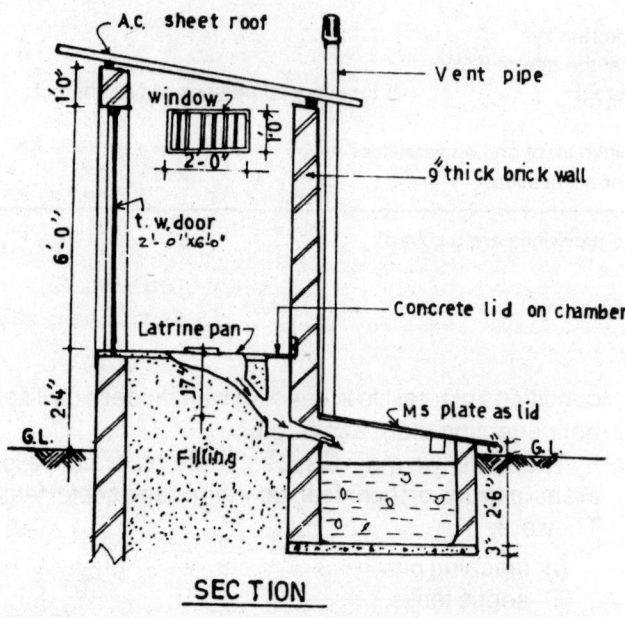

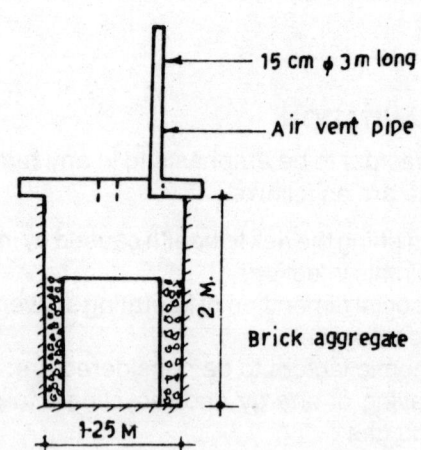

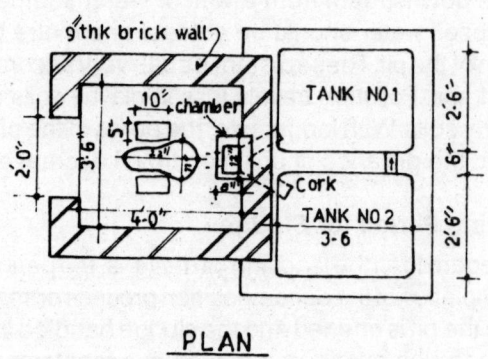

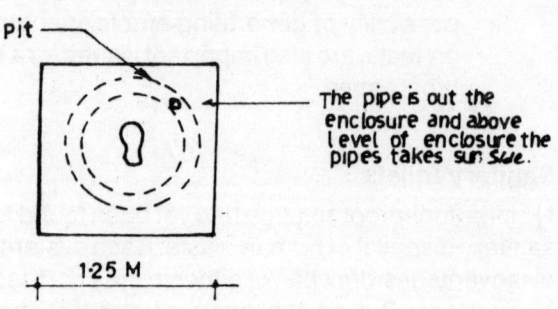

Fig. 16.3 *Plans and sections of "SOPA SANDAS" a ventilated improved double pit latrine.*

Fig. 16.4 *Plans and section of VIP latrine (ventilated improved pit)*

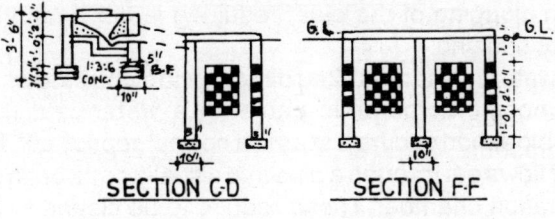

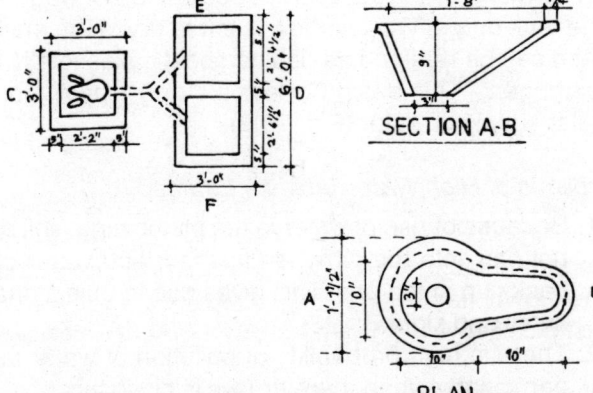

Fig. 16.5 *Details of two pit leaching toilet with water seal trap.*

Two-pit Leaching Toilet with Water-seal Trap (Fig. 16.5)

The double-pit leaching toilet was developed in the late fifties. In this model, a mosaic pan with a P-shaped water-seal trap is attached by a pipe to two displaced pits of 1 metre diameter and 1 metre depth, which are used in turn. The pits can be square or round, lined or unlined according to soil conditions and other factors of convenience.

This design (Fig. 16.6) has been selected as the best low-cost option for the programme of conversion of dry latrines. This type of latrine presupposes low-level subsoil water and availability of sufficient water for flushing.

Pour Flush Water-seal Single-pit Privy

The hard clayey soil with low water table makes a deep unlined pit viable, and it can function for 5 to 7 years before cleaning out is required (Fig. 16.7).

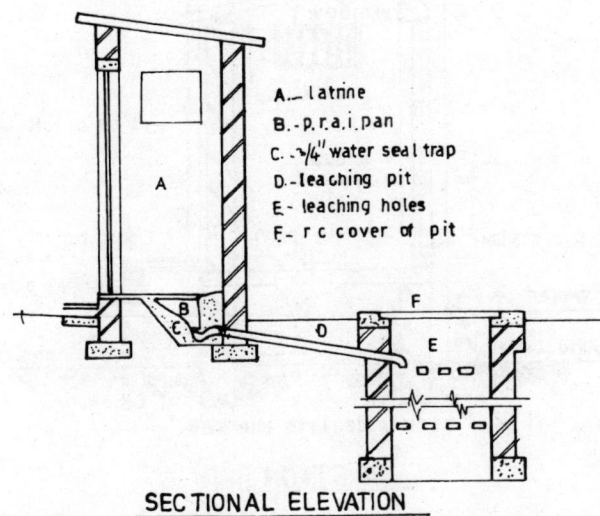

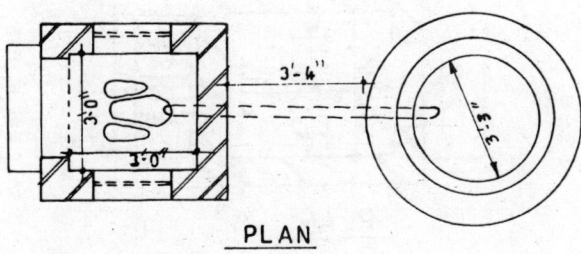

Fig. 16.6 *Detail of leaching pit toilet with water seal trap.*

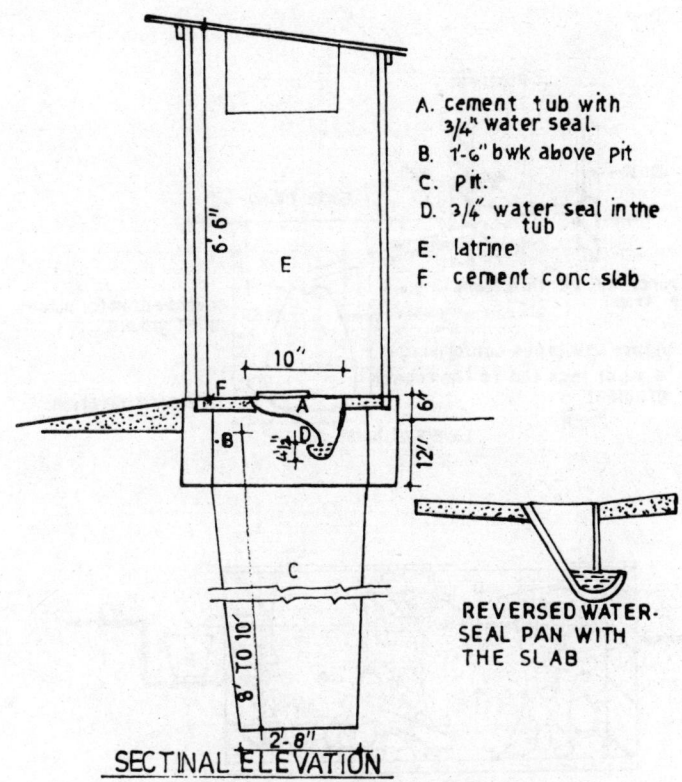

Fig. 16.7 *Detail of pour flush water seal single pit privy.*

Ulta-Matka Privy (Fig. 16.8)

The *ulta-matka* design is the use of a large earthenware pot buried upside down in a pit and serving as its lining. Maximum leaching takes place from the open end at the bottom and only a little from the porous sides of the *matka*. It should be lined with horse dung in the beginning to accelerate the decomposition process. The one-month sealing period before the pit is opened and emptied for reuse is sufficient to destroy all pathogens.

Aqua-Privy

In this type of latrine, anaerobic decomposition takes place in a watertight tank. The solids settle at the bottom in the form of sludge, while the effluent flows out into a soakpit, a water seal is maintained by the drop pipe from the WC pan being submerged about 4 inches below the water level in the tank. This system presupposes availability of a certain amount of water. The drop pipe is slightly curved so that faecal matter is not visible from the top. However, this can result in clogging of the pipe, requiring larger amount of water for flushing.

The watertight tank is sited directly under the seat which has a mosaic or ceramic pan with a water-seal trap. Anaerobic action occurs just as in a normal septic tank. The effluent flows out through a pipe into a soakage pit or an up-flow filtration chamber. These require to be cleaned after two years and six months respectively. The septic tank is connected by an overflow pipe to an adjoining manure pit. After six months, this outlet is opened and the sewage level in the tank drops. The overflow is left to dry in the manure pit and can be retained as rich compost cakes totally free of pathogens. This design is feasible only where there is regular supply of water (Fig. 16.9).

Problems of High Water Table for Latrines

1. Because of rise of water in the pit the night soil does not flow into the pit when flushed, and water often backs up in the pan. This gives rise to objectionable smell and sight.
2. There is high probability of pollution of water table, particularly when there are wells close by.
3. There is a danger of collapse of the pit when the water table rises high and the soil is loose.

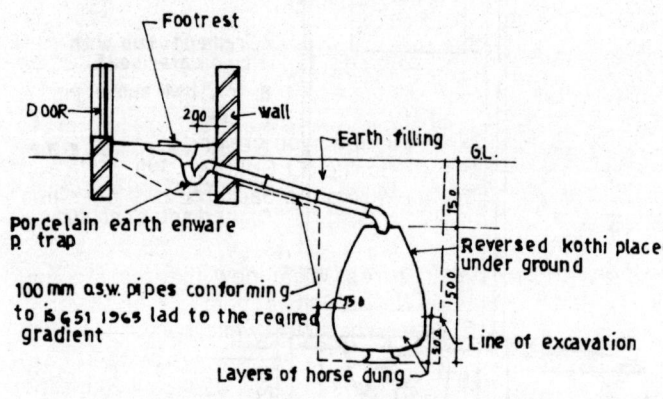

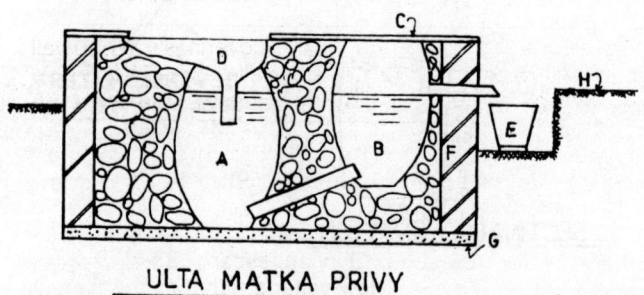

ULTA MATKA PRIVY

A. Grit chamber
B. Septic chamber
C. Removable stone slab
D. Pan with drop pipe
E. Bucket for carrying effluent to manure pit
F. 9" brick wall in mud mortar
G. 6th lime conc.
H. Ground level.

Fig. 16.8 *Detail of ulta-matka privy.*

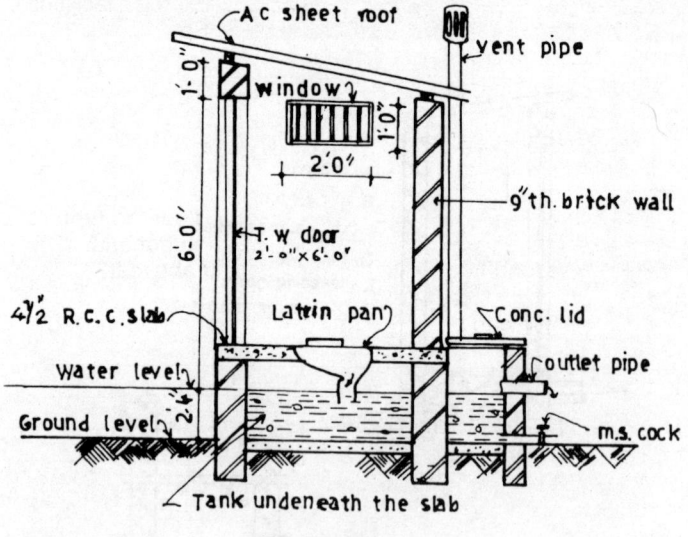

SECTION

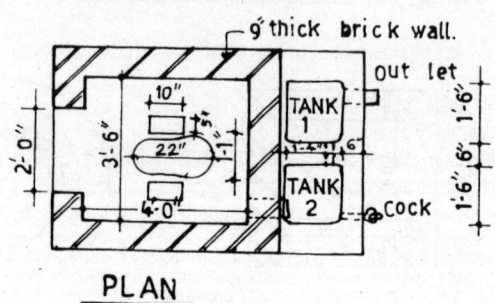

PLAN

Fig. 16.9 *Detail of aqua privy.*

APPENDICES

Appendix A

A BATHROOM QUESTIONNAIRE

This questionnaire will help in critically analyzing the problem and understanding the present status. It contains the detailed information regarding one's liking & disliking and detailed information about the requirement.

1. What's your main reason for changing your bathroom?

2. How many people will be using the room? List adults, children, and their ages.

3. Are users left-handed? Right-handed? How tall is each one?

4. How many other bathrooms do you have?

5. What secondary activity areas would you like to include?
☐ Desk ☐ Garden ☐ Laundry facilities ☐ Exercise facilities
☐ Dressing or makeup area ☐ Sauna ☐ Spa

6. Are you planning any structural changes?
☐ Room addition to existing house
☐ Greenhouse window or sunroom
☐ Skylight ☐ Other

7. Is the bath located on the first or second floor? Is there a full basement, crawl space, or concrete slab beneath it? Is there a second floor, attic, or open ceiling above it?

8. If necessary, can present doors and windows be moved?

9. Do you want an open or vaulted ceiling?

10. What's the rating of your electrical service?

11. What type of heating system do you have? Does any ducting run through a bathroom wall?

12. Is the bath to be used by a physically challenged person? Is the individual confined to a wheelchair?

13. What is the style of your house's exterior?

14. What style (for example, high-tech, country contemporary, country French) would you like for your bathroom? Do you favour compartmentalized European layouts or a more open, informal look?

15. What color combinations do you like?

16. What cabinet material do you prefer: wood, laminate, or other? If wood, should it be painted or stained? Light or dark? If natural, do you want oak, maple, pine, cherry?

17. Storage requirements?
☐ Medicine cabinet ☐ Linen closet ☐ Drawers ☐ Cabinets
☐ Laundry hamper or chute ☐ Rollout baskets ☐ Open shelving ☐ Other

18. What countertop materials do you prefer?
☐ Laminate ☐ Ceramic tile ☐ Solid-surface ☐ Wood
☐ Stone ☐ Other. Do you want a backsplash of the same material?

19. List your present fixtures. What new fixtures are you planning? ☐ Bathtub ☐ Tub/Shower combination ☐ Vanity ☐ Sink ☐ Toilet ☐ Bidet. What finish: white, pastel, full color?

20. Would you prefer natural or mechanical ventilation?

21. What flooring do you have? Do you need new flooring?
☐ Wood ☐ Vinyl ☐ Ceramic tile ☐ Stone ☐ Other

22. What are present wall and ceiling coverings? What wall treatments do you like? ☐ Paint ☐ Wall-paper ☐ Washable vinyl wallpaper ☐ Wood ☐ Faux finish ☐ Ceramic tile ☐ Other

23. Consider natural light sources: ☐ Skylight ☐ Window ☐ Clerestory

24. Artificial lighting desired: ☐ Incandescent ☐ Fluorescent ☐ Halogen ☐ 120-volt or low-voltage. What fixture types? ☐ Recessed downlights ☐ Track lights ☐ Wall-mounted fixtures ☐ Ceiling-mounted fixtures ☐ Indirect soffit lighting

25. What time framework do you have for completion?

26. What budget figure do you have in mind?

Appendix B

COMPARING LIGHT BULBS & TUBES

INCANDESCENT A-bulb	**Description.** Familiar pear shape; frosted or clear. **Uses.** Everyday household use.
T-Tabular	**Description.** Tube shaped, from 5" long, Frosted or clear, **Uses.** Appliances, cabinets, decorative fixtures.
R-Reflector	**Description.** White or silvered coating directs light out end of funnel shaped bulb. **Uses.** In directional fixtures; focuses light where needed.
Par-Parabolic aluminized reflector	**Description.** Similar to Auto headlamp; special shape & coating project light & control beam. **Uses.** In recessed downlights and track fixtures.
Silvered bowl	**Description.** A-bulb, with silvered cap to cut glare and produce indirect light. **Uses.** In track fixtures and pendants.
Low-voltage strip lights	**Description.** Like Christmas tree lights; in strips or tracks, or encased in flexible, waterproof plastic. **Uses.** Task light and decoration.
FLOURESCENT Tube	**Description.** Tube-shaped, 5" to 96" long, Needs special fixtures & ballast. **Uses.** Shadowless worklight, also indirect lighting.
PL-Compact tube	**Description.** U-shaped with base; 5-¼" to 7-½" long. **Uses.** In recessed downlights; some PL tubes include ballasts to replace A-bulbs.
QUARTZ HALOGEN High intensity	**Description.** Small, clear bulb with consistently high light output. **Uses.** In specialized task lamps, torcheres, and pendants.
Low-voltage MR-16 (mini-reflector)	**Description.** Tiny (2" diameter) projector bulb; gives small circle of light from a distance. **Uses.** In low-voltage track fixtures, mono-spots and recessed downlights.
Low-voltage PAR	**Description.** Similar to auto headlamp, tiny filament, shape and coating give precise direction. **Uses.** To project a small spot of light a long distance.

Appendix D
PLATES

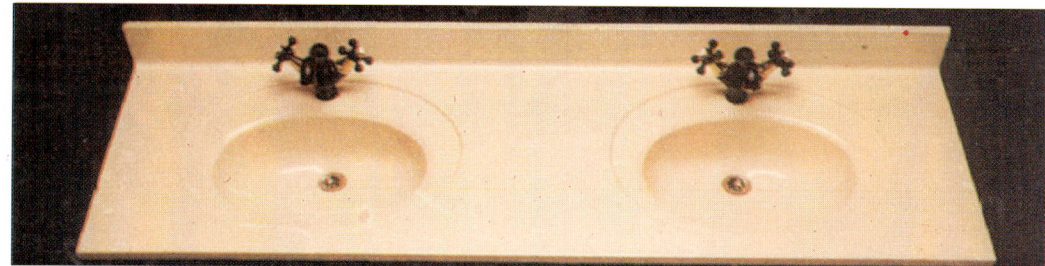

Double Oval: Charmingly sculpted, it draws appreciation from all. The additional basin provides an interesting variation.

Flush in Oval: Built for sheer convenience with smooth sloping sides that help in quick drainage and drying.

Diamond: A contemporary piece, it is artistically multi-faceted. The clean geometric lines help in quick drainage.

Semi Oval: The single piece counter helps save space and avoid splashing.

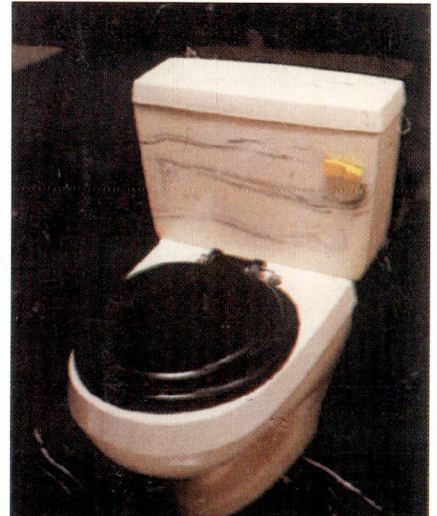

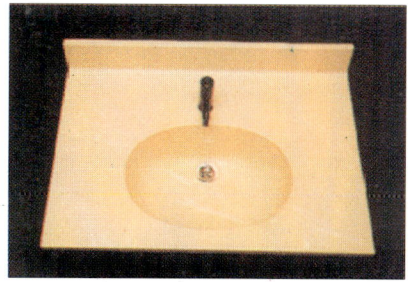

Trapezoid: The eye-catching shape combines sheer elegance and practical convenience.

Plain Oval Mini: Simple and sleek, it provides total economy of space.

EWC
The elegant cast marble water closet is accompanied by a single siphon system, crafted for utmost comfort and hygiene.

Use of cast polymer in washbasin counters, bathtubs, water closets, shower trays, etc.

Appendix C

COST OF BATHROOM

How much will your new bathroom cost? Average bathrooms in United Kingdom cost about 9000 pounds (Rs. 5,00,000). It costs ranging from Rs. 50,000/- to Rs. per Bathroom in India. You may simply need to replace a counter top, add recessed downlights or exchange a worn-out Bathtub to achieve a satisfying change. On the other hand, the sky is the limit Extensive structural changes coupled with ultra-high-end materials and fixtures can easily be added to achieve exclusiveness and richness.

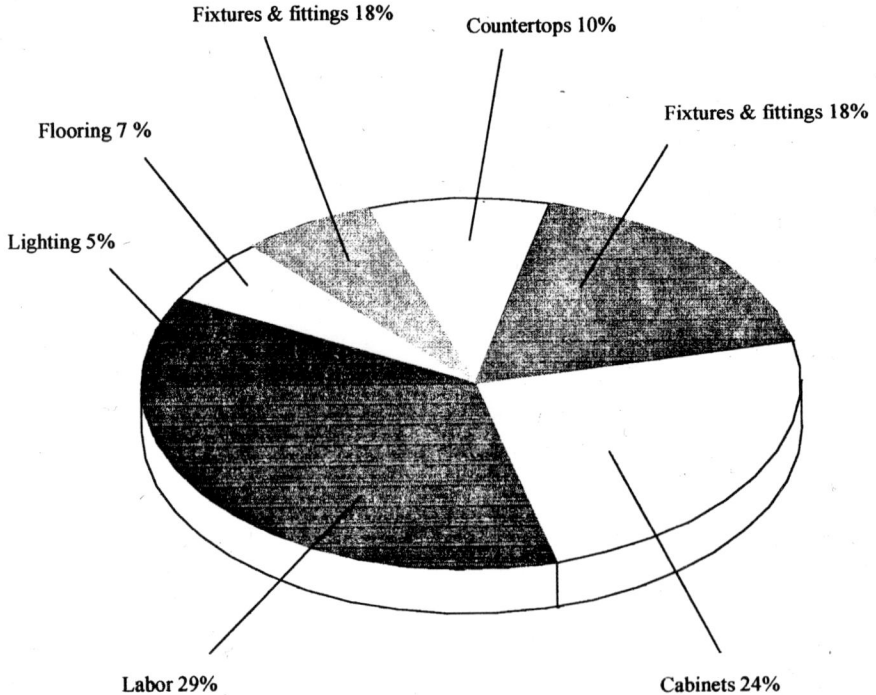

Fixtures & fittings 18%

Countertops 10%

Flooring 7 %

Fixtures & fittings 18%

Lighting 5%

Labor 29%

Cabinets 24%

Don't always make price your only criterion for selection of material. Quality of work, reliability, rapport and on-time performance are some of the very important issues. The percentage cost analysis of various items of Bathroom is explained in the sketch above.

Use of hand made tiles and mouldings as washbasin counter and storage space below washbasin counters.

Appendix E

SANITATION TIPS

With advancement in technology and information systems, the profession of architecture is becoming more complex. It has become rather essential to use the latest innovative construction techniques, materials and construction management systems to evaluate, assess and implement the various projects.

Water supply and sanitation is one such specialized subject, and though very important, is mostly neglected. It has often been seen that a good piece of architecture becomes an eyesore with leaking pipes, choking gully traps and manholes. This happens due to lack of basic knowledge, improper selection of materials, bad detailing and, of course, bad quality workmanship.

Though there are innovative ideas and designs for various products which have enhanced their cosmetic values, the fact remains that the skeleton/structural component for water supply and sanitation needs to be made stronger, effective, and suitable. The various joints for the disposal of waste water from the wash basin, bathing area, kitchen with improper detailing, lead to leakage in the slab, and eventually make the bathroom wet and damp. Therefore, the use of multi-floor traps should be advocated and the plumbers should be educated how to avoid unhealthy leakages.

Most of the time the actual end-user and also the professionals are dependent on the plumbers. However, the plumber with little knowledge, blindly follows the conventional systems without any know-how of the latest materials and their applications. This is how problems multiply, leading to bad detailing and bad workmanship. We still adopt *soot gola* and *safeda* with conventional methods of threading in all our GI pipes, whereas technology has actually changed and we now have revolutionary and wonderful joinery chemicals and tapes that can be used for maintaining better quality of workmanship.

Most of the fittings nowadays being designed and manufactured are having sophisticated sensors, may it be a hand-drier, the flushing of the urinals or washbasin taps, etc. But these high quality fittings, need high quality installation also. The life of the product is reciprocal to the proper use of the product. This is only possible through proper awareness generation among the professionals and plumbers. The industry should arrange some training and orientation programmes for plumbers to sense the product and also understand its acute intensity and behaviour in respect to its high-tech quality and application. Water supply and sanitation is a complex and complicated aspect of the building industry which invariably in our country, is absolutely neglected, and one has yet to see any building which does not have any inbuilt plumbing defects. The same is apparent in the overflow of the water tanks, leakage in the floor traps, dampness in the sunken portion of the bathroom, and blocked, choked gully traps and manholes.

The availability of water proofing construction compounds is indeed revolutionary. All the sunken portions in the toilets, if properly waterproofed with polyurethane/acrylic-based compounds, would avoid such leakages. Similarly when we design hot cold water systems in our bathrooms, there should be a proper insulation of the host water pipes connected to electric geysers or solar geysers. This would avoid heat loss in water travelling from the source to the tap. The hardness of water also plays a very vital and important role in determining the behaviour pattern of various materials. Therefore, if possible, proper treatment should be provided to the source water by way of chlorinization, etc. The uninterrupted water supply through the use of auto-lift pumps, having sensors in underground and overhead tanks are very effective and useful.

The availability of sanitary, chinaware fittings in respect to water closets with high/low level systems or flushing valves are the various factors where a proper selection is required. Similarly, the lack of knowledge pertaining to drainage in most of the projects and in city and town planning creates a lot of problems. Providing of proper slopes, gully gratings, use of stoneware, RCC spun pipes, etc., with proper joints has to be analyzed. Starting from the basic layouts to the various applications of the materials, selection of any booster/injecting pumps, making of bore-wells, etc., a conscious effort to reduce practical problems should be made by professionals in consultation with various manufacturers and the end-users. Unless and until we change our style of functioning and implementation, no change will happen.

Alternative materials like PVC, copper pipes, etc. should be used in the building industry. These are definitely better than the conventional materials which are being used through the ages. Worldover, PVC pipes and fittings have replaced the CI pipes, but here in India we still follow the conventional and do not go for any change. And 'change' is required for a healthy environment of tomorrow.

This article, written by the author, has appeared in *Architecture + Design (A + D)*, March–April 2001.

APPENDICES

Appendix A

A BATHROOM QUESTIONNAIRE

This questionnaire will help in critically analyzing the problem and understanding the present status. It contains the detailed information regarding one's liking & disliking and detailed information about the requirement.

1. What's your main reason for changing your bathroom?

2. How many people will be using the room? List adults, children, and their ages.

3. Are users left-handed? Right-handed? How tall is each one?

4. How many other bathrooms do you have?

5. What secondary activity areas would you like to include?
   □ Desk □ Garden □ Laundry facilities □ Exercise facilities □ Dressing or makeup area □ Sauna □ Spa

6. Are you planning any structural changes?
   □ Room addition to existing house
   □ Greenhouse window or sunroom
   □ Skylight □ Other

7. Is the bath located on the first or second floor? Is there a full basement, crawl space, or concrete slab beneath it? Is there a second floor, attic, or open ceiling above it?

8. If necessary, can present doors and windows be moved?

9. Do you want an open or vaulted ceiling?

10. What's the rating of your electrical service?

11. What type of heating system do you have? Does any ducting run through a bathroom wall?

12. Is the bath to be used by a physically challenged person? Is the individual confined to a wheelchair?

13. What is the style of your house's exterior?

14. What style (for example, high-tech, country contemporary, country French) would you like for your bathroom? Do you favour compartmentalized European layouts or a more open, informal look?

15. What color combinations do you like?

16. What cabinet material do you prefer: wood, laminate, or other? If wood, should it be painted or stained? Light or dark? If natural, do you want oak, maple, pine, cherry?

17. Storage requirements?
    □ Medicine cabinet □ Linen closet □ Drawers □ Cabinets □ Laundry hamper or chute □ Rollout baskets □ Open shelving □ Other

18. What countertop materials do you prefer?
    □ Laminate □ Ceramic tile □ Solid-surface □ Wood □ Stone □ Other. Do you want a backsplash of the same material?

19. List your present fixtures. What new fixtures are you planning? □ Bathtub □ Tub/Shower combination □ Vanity □ Sink □ Toilet □ Bidet. What finish: white, pastel, full color?

20. Would you prefer natural or mechanical ventilation?

21. What flooring do you have? Do you need new flooring?
    □ Wood □ Vinyl □ Ceramic tile □ Stone □ Other

22. What are present wall and ceiling coverings? What wall treatments do you like? □ Paint □ Wall-paper □ Washable vinyl wallpaper □ Wood □ Faux finish □ Ceramic tile □ Other

23. Consider natural light sources: □ Skylight □ Window □ Clerestory

24. Artificial lighting desired: □ Incandescent □ Fluorescent □ Halogen □ 120-volt or low-voltage. What fixture types? □ Recessed downlights □ Track lights □ Wall-mounted fixtures □ Ceiling-mountecl fixtures □ Indirect soffit lighting

25. What time framework do you have for completion?

26. What budget figure do you have in mind?

Appendix B

COMPARING LIGHT BULBS & TUBES

INCANDESCENT A-bulb	**Description.** Familiar pear shape; frosted or clear. **Uses.** Everyday household use.
T-Tabular	**Description.** Tube shaped, from 5" long, Frosted or clear, **Uses.** Appliances, cabinets, decorative fixtures.
R-Reflector	**Description.** White or silvered coating directs light out end of funnel shaped bulb. **Uses.** In directional fixtures; focuses light where needed.
Par-Parabolic aluminized reflector	**Description.** Similar to Auto headlamp; special shape & coating project light & control beam. **Uses.** In recessed downlights and track fixtures.
Silvered bowl	**Description.** A-bulb, with silvered cap to cut glare and produce indirect light. **Uses.** In track fixtures and pendants.
Low-voltage strip lights	**Description.** Like Christmas tree lights; in strips or tracks, or encased in flexible, waterproof plastic. **Uses.** Task light and decoration.
FLOURESCENT Tube	**Description.** Tube-shaped, 5" to 96" long, Needs special fixtures & ballast. **Uses.** Shadowless worklight, also indirect lighting.
PL-Compact tube	**Description.** U-shaped with base; 5-¼" to 7-½" long. **Uses.** In recessed downlights; some PL tubes include ballasts to replace A-bulbs.
QUARTZ HALOGEN High intensity	**Description.** Small, clear bulb with consistently high light output. **Uses.** In specialized task lamps, torcheres, and pendants.
Low-voltage MR-16 (mini-reflector)	**Description.** Tiny (2" diameter) projector bulb; gives small circle of light from a distance. **Uses.** In low-voltage track fixtures, mono-spots and recessed downlights.
Low-voltage PAR	**Description.** Similar to auto headlamp, tiny filament, shape and coating give precise direction. **Uses.** To project a small spot of light a long distance.